WELCOME

In October 1957, the world held its breath when it learned that the Soviet Union had launched the world's first artificial satellite. In the midst of the Cold War, it was perceived as a threat to global peace and demonstrated that Russia had big rockets that could strike the United States with atomic weapons.

Less than four months later, America put its own satellite into orbit and the Space Race began, a contest for national pride and technical capability that only ended with the collapse of the Soviet Union in 1991. In those intervening years, public attention turned to the race to get a human in space, carry out a rendezvous and docking in orbit, launch increasingly more powerful rockets and put men on the Moon.

Less noticed, along the way, a wide range of satellites were developed for diverse applications, including spying over contested places, remote sensing of the Earth and its resources, managing global communications, observing, and predicting weather and building a worldwide navigation system for public and private use. Satellites were the real story behind the race for space – less dramatic but infinitely more important for the everyday lives of ordinary individuals.

Today, in the 21st century, we switch on to watch news broadcasts and sporting events from the other side of the world, get weather forecasts about places we plan to be visiting in a few hours or days, receive messages from friends and relatives in far-off places and get maps and location information from smartphones and smartwatches – all enabled by satellites. And when we are in danger, we can get first responders to remote and inaccessible places devoid of conventional communications.

Behind these day-to-day advantages we all take for granted, satellites provide information building a bigger picture about our environment, our natural resources, and the climate now and predicted in the coming decades. This holistic view of our home planet and the ensuing connections between people ensures a greater access to information, knowledge, opinion, and analysis than humans have ever had before. And it comes as a new and expanding industry stretches across every country on Earth.

Today, the space programme is worth $460 billion a year in jobs, products, commodities, services, entertainment, efficiency-savings, and development of sustainable resources. It is the fastest growing industry on the planet, more than a million people getting pay checks through companies, organisations and government departments employed directly or indirectly by space projects and programmes.

Where once the global space industry was taxpayer funded, the vast majority of investment now comes from non-government bodies and organisations. Of national agencies, the US government's civilian space programme managed by NASA accounts for almost half the world total. Yet it receives less than 0.4 per cent of total annual national government expenditure each year and puts back four times that much into the American economy.

The story of the rise and rise of satellite services is both a history lesson and a catalogue of services and opportunities created and energised by an industry which did not exist 65 years ago at the dawn of the Space Age. In broader context, the space programme has enabled us to know more about our Earth than we ever did before, to take its pulse with a constellation of diagnostic sensors electronically plugged in to the land, sea, and air so that we know the truth behind our changing environment.

In this book we scratch the surface and provide a window on the unsung benefits of the space programme, peering behind dominating cover-images of derring-do on distant worlds. To bring home the reality of a world of nations made interdependent as never before by unseen webs of information binding humans together on a busy and crowded planet.

David Baker
Editor

(NASA)

(Boeing)

(Author's collection)

CONTENTS

(Author's collection)

(Boeing)

(ESA)

ISBN: 978 1 80282 784 2
Editor: David Baker
Senior editor, specials: Roger Mortimer
Email: roger.mortimer@keypublishing.com
Cover design: Steve Donovan
Design: SJmagic DESIGN SERVICES, India
Advertising Sales Manager: Brodie Baxter
Email: brodie.baxter@keypublishing.com
Tel: 01780 755131
Advertising Production: Debi McGowan
Email: debi.mcgowan@keypublishing.com

SUBSCRIPTION/MAIL ORDER
Key Publishing Ltd, PO Box 300, Stamford, Lincs, PE9 1NA
Tel: 01780 480404
Subscriptions email: subs@keypublishing.com
Mail Order email: orders@keypublishing.com
Website: www.keypublishing.com/shop

PUBLISHING
Group CEO and Publisher: Adrian Cox
Published by
Key Publishing Ltd, PO Box 100, Stamford, Lincs, PE9 1XQ
Tel: 01780 755131 **Website:** www.keypublishing.com

PRINTING
Precision Colour Printing Ltd, Haldane, Halesfield 1, Telford, Shropshire. TF7 4QQ

DISTRIBUTION
Seymour Distribution Ltd, 2 Poultry Avenue, London, EC1A 9PU
Enquiries Line: 02074 294000.

WHY GO TO SPACE?

The challenge of crossing new frontiers plus the excitement of doing something new proves irresistible.

The desire to explore space probably came from a combination of the human instinct to find the highest hill, climb the highest mountain and sail to distant lands, mixed in with the challenge of identifying a new frontier and finding the means to cross it. Added to this is the seductive desire to tackle very difficult tasks so as to enhance the sense of fulfilment and personal reward. More so if the solution is hard to find with satisfaction never greater than the challenge of getting to hitherto unexplored places.

Then there is the sheer excitement of doing something daring, risking all in the pursuit of adventure and living the dream. Writers and futurists have been thinking and talking about it for centuries, stepping off the Earth and flying to other worlds, visiting other planets, exploring solar systems around distant stars and colonising alien places in the deep recesses of endless space. It was all out there to be viewed every night, beckoning the adventurous spirit, and charging the adrenalin rush with new energy. Whether to the Moon or Mars, the pull of the unknown became stronger with time.

RIGHT • An enduring exponent of new and exciting scientific innovations, H G Wells (1866-1946) foretold through his imaginative writings the time when humans would go into space. (George Charles Beresford)

RIGHT • The French novelist Jules Verne (1828-1905) wrote about the emerging technologies of the 19th century and created an expectation of adventure and discovery on this and other worlds. (Etienne Carjat)

Neither did it help that there was a surge in science fiction to whet the 19th century appetite for all manner of scientific speculation as to the possible, the plausible or the probable. It was a century of great discoveries, astounding inventions and speculation about flight, rocketry, space travel and visits to other worlds. No surprise therefore that science fiction began to merge with science fact and to speculate on imaginative concepts where, to many dreamers, all things seemed possible. Writers such as Jules Verne and H G Wells fuelled the appetite for speculative drama with predictions of flights to the Moon and of alien civilisations on Mars mobilising for war.

For most of the 19th century, exploration and global expansion was a by-product of empire, with France, The Netherlands, Germany and not least Great Britain colonising distant places on Earth, extracting human and material resources in a great age of industrial expansion. With a grand mobilisation of the machine age came wealth and a stimulus for new technology, the steam engine replacing horsepower, the locomotive replacing the canal barge and the screw-driven cargo ship replacing sail. Soon,

LEFT • Born in Budapest, Theodore von Kármán (1881-1963) studied supersonic and hypersonic airflow and proved an inspiration to a new generation of engineers and lent his name to the defining boundary between the Earth's atmosphere and space. (Author's collection)

the internal combustion engine would shrink countries and aeroplanes would bridge continents. But space? That was a nurtured aspiration – still a far-off dream, but one sustained by an enduring desire to explore.

Bridging worlds

Exploration is an essential part of the human story and has taken people across seas and oceans from continent to continent and from island to island, spreading humans throughout the world over the last 50,000 years. It is a natural progression from land and sea to air and space, a transition that began 120 years ago when the Wright Brothers first took to the skies in December 1903. Little more than 40 years after that, the first flight into space took place on June 20, 1944, when a V2 rocket exceeded what is now regarded as the upper limit of the atmosphere.

As the Earth's atmosphere becomes less dense with altitude, that defining boundary is an arbitrary point, one now generally agreed as a height of 100km (62 miles) above sea level. That is known as the Kármán Line after the Austrian-Hungarian scientist Theodore von Kármán, who defined that boundary as the approximate altitude where aerodynamic flight is maintained by inertia rather than lift. It is also the minimum altitude at which satellites can complete one revolution of the Earth without propulsion – an orbit. But it is vague and arbitrary since there is no physical or aerodynamic value other than those expressed by von Kármán himself. And it is a nice, rounded number!

That V2 rocket flight achieved a maximum altitude of 174.6km (108.5 miles) before falling back to Earth but it was a portent of things to come and demonstrated the potential in the new field of liquid propellant rocketry. Major gains in this technology had been achieved by the German Army after it appointed Walter Dornberger in 1930 to head a programme under the direction of civilian rocket scientist Dr Wernher von Braun. Their pioneering work was applied to a ballistic missile with a range of approximately 322km (200 miles) bearing the Army designation A-4 but known colloquially as the V2, for Vergeltungswaffe (revenge weapon). The V1 was a flying bomb powered by a pulsejet engine which flew to its target at subsonic speed, whereas the rocket powered V2 had a maximum speed of 5,760kph (3,580mph).

Getting into orbit was achievable using liquid propellant rocket motors rather than solid propellant types which had been around for several centuries. Solid propellants are a mix of fuel and oxidiser so that when ignited they burn at high temperature and expel exhausted products through a shaped exit nozzle. That provides the action from which a reaction propels the missile forward. Liquid propellants have a much greater energy per unit mass of propellant and can achieve higher exhaust velocities, a key to engine performance. However, solid rocket motors are far simpler than liquid engines and generally cheaper to produce, although complexity and sheer size can reverse that latter advantage.

Solid propellant rocket motors are storable and can be held for very long periods awaiting use, whereas optimised liquid propellants are not storable for more than a few hours and cannot be held in perpetual readiness for flight. There are exceptions, such as storable liquid propellants, but these do not have the same efficiency and performance as those using super-cold (cryogenic) propellants such as liquid oxygen as oxidiser and liquid hydrogen as fuel. Perhaps not surprisingly, it was the combination of these two liquid propellants that fired up enthusiasm for space travel with a scientifically viable solution to the need for speed and high performance to leave Earth's atmosphere.

Extraordinary steps

Throughout the 19th century, science and engineering achieved extraordinary steps, making possible new technologies in chemistry and materials, many of which began to make space travel appear feasible. The Russian teacher Konstantin Tsiolkovsky (1857-1935) joined a growing band of intellectuals, scientists, engineers, and philosophers in the Russian cosmism movement, advocating interplanetary travel and flights to other worlds. Great progress was being made in understanding the scale and complexity of the Universe and this attracted theorists and futurists to merge discovery with the desire to explore the planets.

Tsiolkovsky studied the use of liquid hydrogen and liquid oxygen as highly efficient propellants for getting off the

LEFT • The Russian school teacher Konstantin Tsiolkovsky (1857-1935) provided the scientific foundation that stimulated Russian engineers to assemble a base of research from which came their great progress in astronautics. (Author's collection)

RIGHT • Dr Robert Goddard worked on the development of liquid propellant rockets in the 1930s, attracting little interest from fellow engineers or government. (Author's collection)

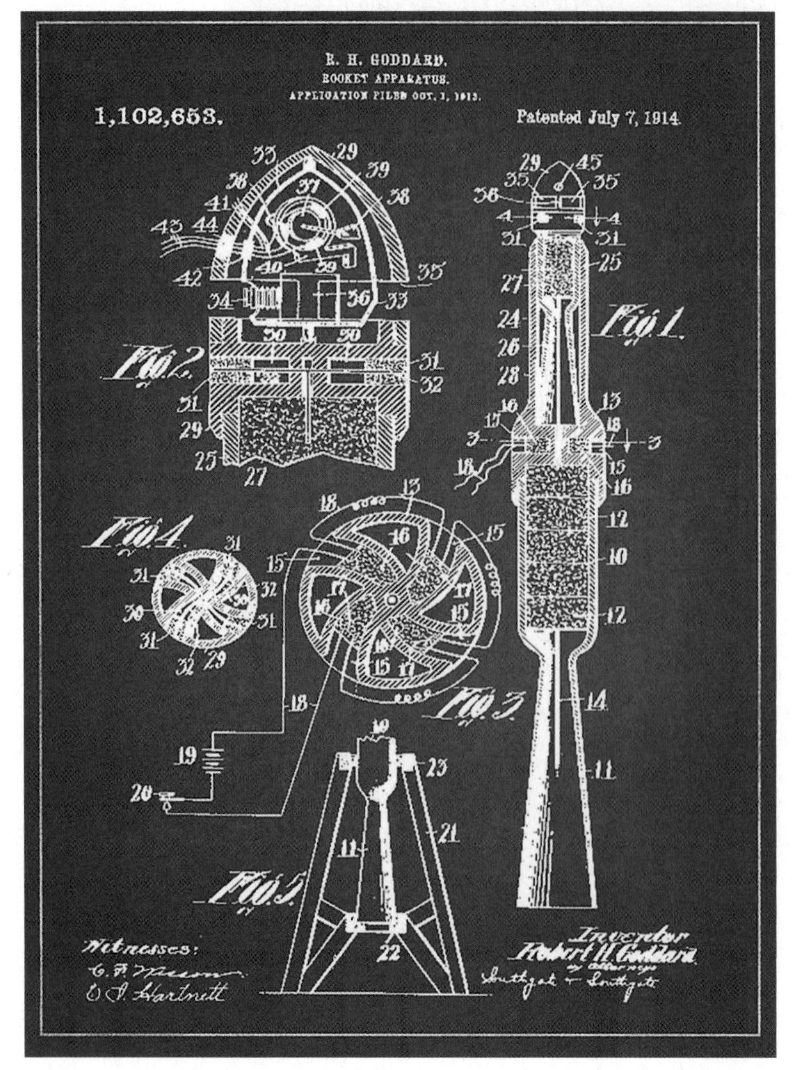

BELOW • A patent paper from some of Robert Goddard's work on both solid and liquid propellant rockets. (Author's collection)

Earth and into space. As the lightest element, hydrogen was already being used for lighter-than-air dirigibles, airships which would be greatly exploited for travel and transport before the First World War (1914-1918). It would be many decades before those propellants were harnessed for rocket propulsion, but Tsiolkovsky's ground-breaking ideas stimulated a growing body of supporters in Russia for interplanetary space travel.

While the theoretical possibilities were developed and refined in the emerging Soviet Union, the first liquid propellant rocket was launched by the American engineer and inventor Robert H Goddard (1882-1945) on March 16, 1926. This opened the door on small research projects funded by private groups of enthusiasts and by scientists and engineers working within funded laboratories. But in the Soviet Union, a major government initiative was supported for a broad range of research and test projects. Nevertheless, in addition to the Soviet Union, it was in the United States that progress was made with experimental liquid propellant rockets.

In Germany, Hermann Oberth wrote 'The Rocket into Planetary Space', explaining how rockets could be used for scientific research and is generally considered one of the greatest protagonists of space flight. In Britain, and in France especially, groups formed to take this work and conduct their own research, a group in the UK setting up the British Interplanetary Society in 1933 and which still exists, the oldest continuously running space advocacy group in the world. At the same time in the USA, the American Interplanetary Society was formed, quickly renaming itself the American Rocket Society. In France, Esnault-Peltérie conducted research into rocket motor design and that work contributed to a post-war surge in the development of rockets of all types and, in particular, for the scientific exploration of the stratosphere.

The end of World War Two in 1945 saw the gradual restoration of international teamwork for scientific research into the upper atmosphere and near-Earth space. This would lead to the International Geophysical Year (IGY) of July 1, 1957, to December 30, 1958, which followed the tradition of the International Polar Year of 1882-83 and that of 1932-33. Following the turbulent years of political unrest and war, it was a reaffirmation of the truly international nature of scientific research and exploration and in advancing the body of knowledge about the planet by land, sea, and air. The IGY is largely remembered today because it sponsored development of artificial satellites, with Russia being the first to launch an object into orbit when it put Sputnik 1 up on October 4, 1957, followed by the Americans with Explorer I on January 31, 1958.

Both countries had been sending rockets into space for some time – 'sounding-rockets' as they were called – carrying packages of instruments, most of which were recovered by parachute from high in the atmosphere. More than 200 sounding rockets were launched during the IGY. Inspired by the V2, while the military development of the rocket had been pursued after the war as a priority, it was with captured V2 rockets and German engineers that scientific research into the upper atmosphere and near-Earth space was undertaken by both the Soviet Union and the USA. This largely forgotten start to the Space Age provided the basis upon which the scientific exploration of space began.

Action and reaction

Both Tsiolkovsky and Goddard had each developed the rocket equation based on their own calculations but only Goddard would take it further to conduct experiments with real rockets. Before them, the British mathematician William Moore had calculated the rocket equation as early as 1810, with results published three years later. Defined through calculus, Tsiolkovsky had independently derived the differential equation and published his own treatise in 1903. It explained the ability of a rocket to move forward as a reaction to the action of expelling a portion of its mass as propellant. It also followed that a rocket would achieve acceleration by reducing the mass but maintaining thrust output. In this way, it conforms to the conservation of momentum which had been defined by Isaac Newton in 1687.

For those who prefer a practical explanation, the rocket works by throwing mass out at one end (the action), allowing the main body containing propellant to move in the opposite direction (the reaction). Unlike a gun or an artillery piece, the rocket *is* the reaction rather than the action – providing a sustained recoil as long as propellant is being consumed. The rocket equation was profound, again calculated and independently rewritten by Goddard in 1912 and Oberth around 1920. Thus, armed with a mathematically legitimate way of travelling through space, innovators, engineers and experimenters in Russia, Europe and the United States began working up ways to get off the surface of the Earth and out of the atmosphere.

By definition, rockets are optimised for operating in a vacuum because they combine both fuel and oxidiser and are therefore independent of their external environment. This is unlike the jet engine, also a reaction device but which takes in oxygen from the surrounding air to combust the fuel. Rocket motors are also more efficient in space and not when climbing through the atmosphere, because in the vacuum there is no air resistance (drag) to fight against.

Getting to space is constrained by physical factors impossible to avoid. However brilliant the engineers might be, they have to work within these immutable constraints. These are the fixed gravitational constant for getting off the Earth and the limited potential in the chemical properties of propellant. These constraints make it possible to launch rockets to great altitude but to achieve sustained flight in space, a second rocket on top is needed. An upper stage can take advantage of the speed and height achieved by the first of a two-stage propulsion system by adding its own performance to that of the lower segment.

Multi-stage rockets had been widely proposed from the earliest days of space exploration but the perfection of carrying and igniting an upper stage was considered problematical. Today, multi-stage rockets and space launch vehicles are standard and commonplace but in the

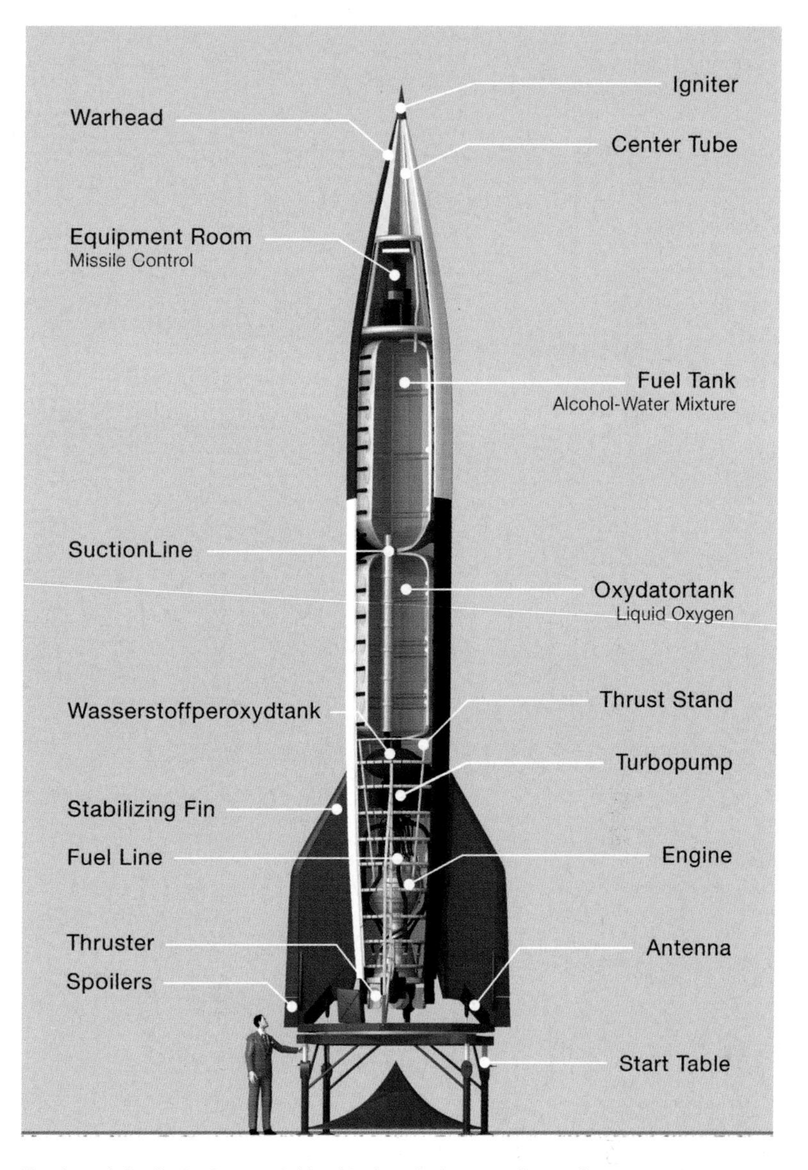

ABOVE • *The first missile to enter space on a ballistic flight was the German A-4, or V2 developed in Germany during the Second World War. (Eberhard Marx)*

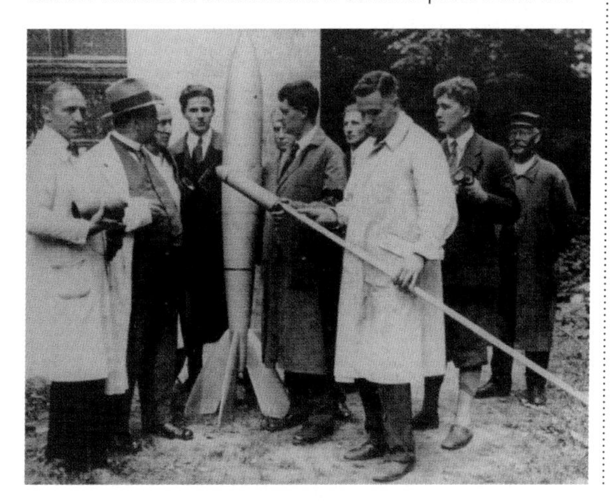

LEFT • *While support for rocketry in the US languished, in Germany there was encouragement from the army. Here, von Braun (second from right) attends a gathering to conduct tests. (Author's collection)*

ABOVE • *In the final year of the Second World War, the German army mobilised rocket troops to deploy and fire several thousand V2 rockets from the occupied Low Countries and France. (Author's collection)*

immediate post-war years they were a great technological challenge. It was proving difficult enough to get a rocket off the ground first time and on time. To add further complexity to the operation of an upper stage required a high degree of reliability and a means of controlling the sequence of events, either by radio signal and remote control from the ground or through internal control systems.

Curiosity and expediency

The desire for detailed information about the upper atmosphere and near-Earth space was driven both by curiosity and by expediency. Aircraft were flying at increasing altitude and higher speeds and to achieve an edge over a potential aggressor, there was increased awareness that a new generation of combat aircraft would evolve rapidly. For that to happen, it was necessary to understand the physics of the new environment in which they would operate and to provide information to build the wind tunnels of the future. Much of that was in the hands of the National Advisory Committee for Aeronautics (NACA) which, since it was formed in 1915, conducted flight research.

RIGHT • *Hermann Oberth (foreground) with (left to right) Ernst Stuhlinger, Maj Gen H N Toftoy (responsible for bringing German scientists to the US), Wernher von Braun and Robert Lusser. (US Army)*

By the end of the war in 1945, aircraft were already being designed to fly at supersonic speeds and achieve altitudes impossible with piston-powered engines. It was vital to obtain data for these about the dynamics of the upper atmosphere. The ready availability of the V2 as a platform for research into rocket design and for carrying scientific instruments was self-evident. With a thrust of 249kN (56,000lb) it was the most powerful rocket on the planet and could reach a maximum altitude of 214km (133 miles). But it was only one tool used for upper atmosphere research.

Universal interest in acquiring information about the extraordinary investment by the German government in aircraft, rockets, and aerodynamics in particular led Britain to launch the first V2 rockets after hostilities. Under the Backfire programme and launched from the island of Cuxhaven with help from German engineers, the British Army conducted its first flight with this missile on October 2, 1945. It reached a height of 69.4km (43.1 miles) and a range of 249.4km (155 miles). It was followed by the second two days later but that failed, the final British launch occurring on October 15, 1945, which repeated the performance of the first with a landing in the North Sea. All three remain on the sea floor to this day.

The first V2 fired from US soil was launched on April 16, 1946, but with determination to employ it as a sounding-rocket for scientific research and by September 19, 1952, the US Army had sent up 78 rockets of this type. During this programme, the tests transitioned to Cape, where the Army and the Navy were establishing facilities for test launches. Along with the rapidly evolving technology, several groups came together to form the Upper Atmosphere Research Panel, which was organised at Princeton University on February 27, 1946. The committee thus formed was to manage activities which included military, academic institutions, and leading university interests. From this came a series of science tasks producing an increasing range of instruments to satisfy those requirements.

Under the codename 'Blossom', modifications were made to the warhead section of the V2, expanding the internal volume from 0.453m^3 (16ft^3) to 2.26m^3 (80ft^3), into which a variety of different scientific instruments and biological samples were placed. With a payload weight of 540kg (1,200lb) instruments were developed at a range of locations and delivered to the Army launch site for these rockets located at White Sands Missile Range, where they could be carried on high-altitude ballistic flights. The first living things sent up were fruit flies launched from White Sands on February 20, 1947, for radiation measurements. The rocket reached a height of 109km (68 miles).

First primate

The first primate in space was launched by a V2 on June 14, 1949, when the rhesus monkey Albert II rode the rocket to a height of 134km (83 miles) but died before it could be recovered on the ground. These initial V2 research flights were but one tool in parallel with a concerted effort to gather information about the upper atmosphere from balloon flights, measuring radiation and testing changes in seeds and small organisms. The first balloon in this programme lifted off on November 18, 1947, from Stagg Field at the University of Chicago, Illinois. Over the following years, a wide range of living organisms including insects and small animals were flown on several balloon flights up to 33 hours in duration.

ABOVE • *Miss Baker, a squirrel monkey which was carried on a ballistic flight to space, one in a long line of apes and monkeys used for research into the biological effects of space flight. (US Army)*

The Army controlled most engineering and flight operations and began a fast-track learning process from German technicians brought to the United States. To expand the amount of time spent in space and to achieve higher altitudes, the Army managed a programme codenamed Bumper to integrate the V2 rocket with an upper stage. Developed by the Jet Propulsion Laboratory (JPL) in 1944, the Corporal rocket formed the basis for this where it was renamed WAC Corporal. Some say that WAC stood for 'without active control' but there are other less politically correct appellations! So that it could operate as an atmospheric research tool, the stage was given a parachute for safe recovery in addition to the parachute for the science instruments and their recording devices, which returned to Earth separately.

The liquid propellant WAC Corporal upper stage used red fuming nitric acid (RFNA) as the fuel, a mixture of 84 per cent nitric acid, 13 per cent dinitrogen tetroxide and a small quantity of water. The oxidiser was furfuryl alcohol, a colourless liquid with a faint smell of burning and a bitter taste never to be forgotten! These propellants are hypergolic, meaning they have chemical combinations which ignite on contact and therefore do not require a source of ignition. That simplifies the design but places demand on the need to keep them separate until brought together in the motor's combustion chamber.

Developed by Aerojet from the original Corporal and first into the WAC A rocket, the improved WAC B produced a thrust of 6,672N (1,500lb) for 45secs. Tests with the rocket began in 1945 and on its own it proved capable of achieving a speed of 4,505kph (2,800mph) and an altitude of 30.5km (19 miles). When attached to the V2, the complete two-stage assembly could lift instruments to a maximum speed of 8,045kph (5,000mph) and a height of at least 389km (242 miles).

The programme was named Bumper (from the 'bump' given to the WAC Corporal by the V2) and there would be eight flights in the series. Primary objectives were research into the separation of an upper stage in flight, which required unusually precise stability in the terminal flight of the first stage – the V2, and to measure the effects of high speed and

LEFT • *A Bumper flight launched from White Sands, New Mexico to V2 stage separation at 32.2km (20 miles) and an altitude of 129km (80 miles), achieved by the WAC Corporal upper stage on February 24, 1949. (US Army)*

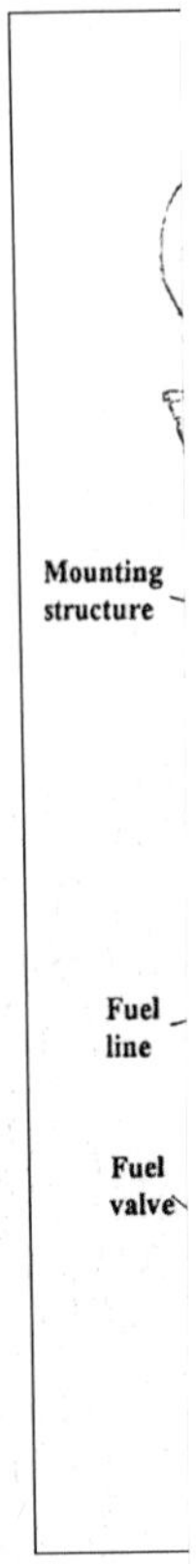

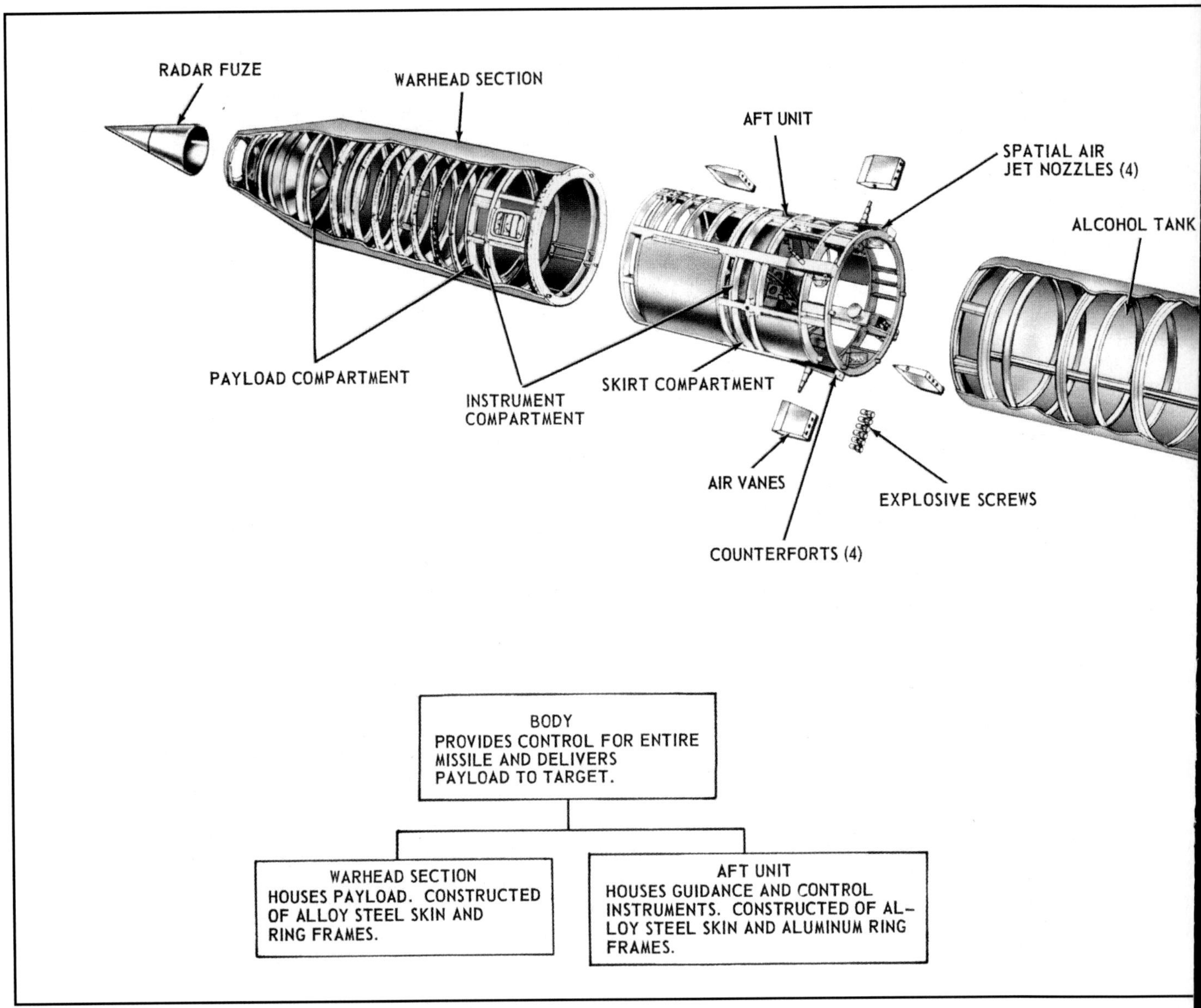

ABOVE • *The Redstone rocket was the first US battlefield missile for theatre operations and was nuclear capable, albeit with very basic movable control fins in the rocket's exhaust for stability and directional control. (Chrysler Corporation)*

1,600kN (360,000lb), or little more than 40 per cent of its Russian contemporary. Requiring non-storable, super-cold liquid oxygen to support ignition of the kerosene fuel, as ICBMs both Atlas and R-7 would be replaced by missiles using storable propellants enabling the deterrent to be instantly ready for launch on demand. While limited as weapons of war, both the R-7 (later named Soyuz) and the Atlas would each have an enduring future as satellite and spacecraft launchers.

The Soviets began flight tests with the R-7 on May 15, 1957, but after successfully getting off its launch pad at Baykonur, Kazakhstan, one of the four boosters caught fire and separated prematurely with the rocket, impacting the ground 400km (248 miles) down-range. The second launch on July 12 was also unsuccessful when the guidance system experienced a malfunction but the third flight on August 21 and the next on September 7 were successful. All four launches were monitored by the US intelligence community from radar sites over the border in Turkey and from Iran, where friendly governments were politically supported by Washington for this and other intelligence gathering purposes. It was with this rocket that analysts in the US believed a satellite would be launched and as the Soviet R-7 flights continued, there was heightened expectation of an attempt.

Development of a Russian satellite had been formally proposed by Korolev on December 17, 1954, just two months after the Americans began discussions about a US satellite for the IGY and seven months before Eisenhower's press secretary James Campbell Hagerty announced what became the Vanguard programme. While the Americans were planning to launch a succession of Vanguard satellites, on January 30, 1956, the Soviet Council of Ministers approved development of a satellite weighing up to 1,400kg (3,087lb). Seven months later, the development began of a two-stage version of the R-7 for sending spacecraft to the Moon and the closest planets Venus and Mars.

Aware that the Americans were pushing ahead with Vanguard, and recognising the limited capacity of that rocket, the Russians were thinking big. By using their powerful ICBM as a space launcher, they could outclass and out-launch the Americans who were relying instead on a small satellite launcher based on an existing sounding-rocket.

Smaller size

But their satellite, Object D, would not be ready before 1958. As early as May 1957, the month in which the second Viking test shot for Vanguard was fired, the Russians began development of a much smaller satellite designated PS-1. They were convinced that this could be rushed into orbit well in advance of the American effort, leaving Object D to come along later.

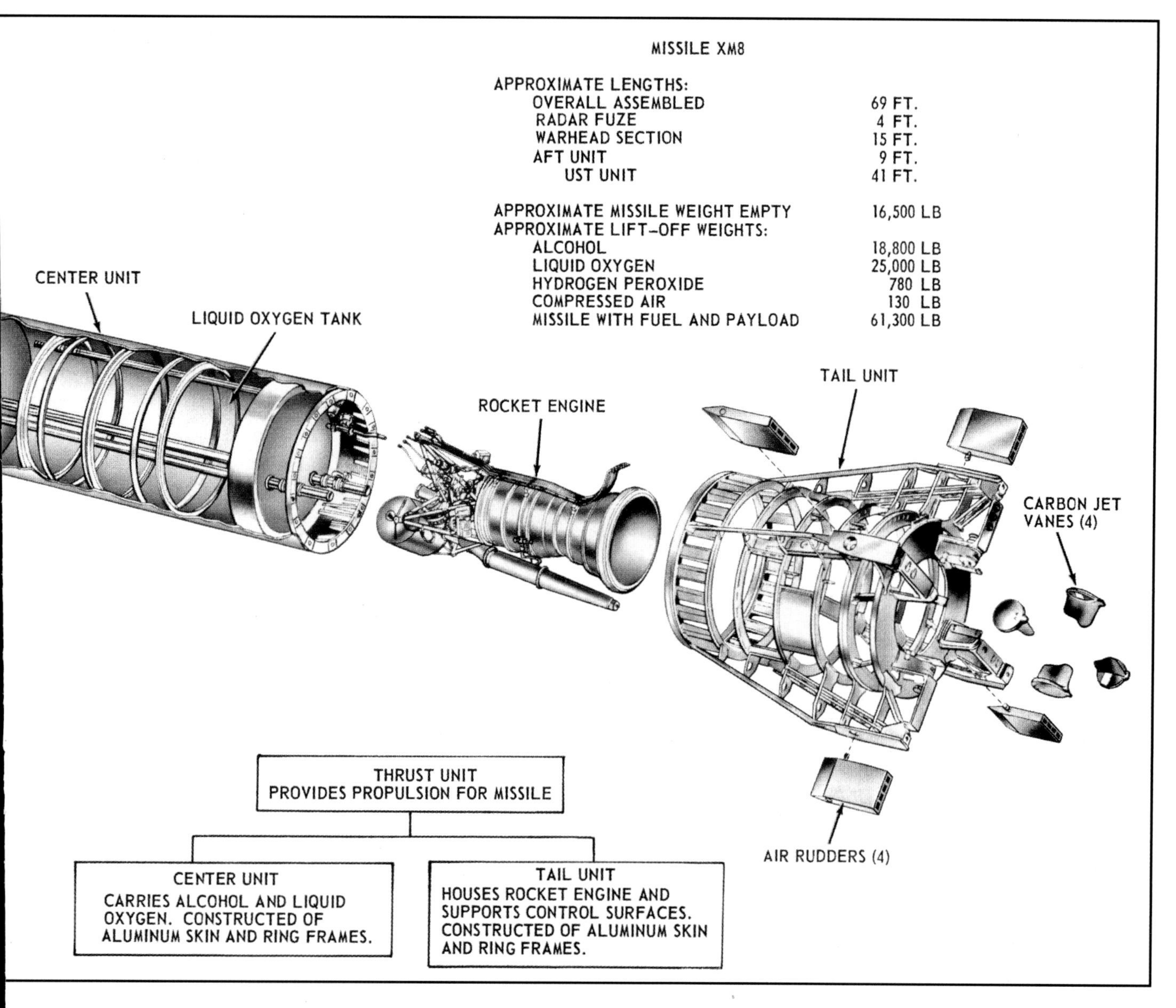

Not unexpected by the US intelligence community, in the early hours of October 5, 1957 local time, the fifth R-7 placed Sputnik 1, the world's first artificial satellite, in an orbit of 936km (583 miles) by 215km (134 miles), making one full circumnavigation of the Earth in 96.2min. It would be forever logged as having been launched on October 4, as measured by the Universal Time Clock (UTC) which had yet to tick over to the following day. Weighing 83.5kg (184lb), PS-1 was a 58cm (23in) diameter sphere trailing four long communication antennae. Power was provided by a single battery and a small transmitter would send signals on frequencies of 20MHz and 40MHz.

After lift-off, the four boosters separated successfully at 1 minute 56 seconds. The core stage continued to fire until 4 minutes 55 seconds and went into orbit. The fairing and the satellite were released together 20 seconds later, all three objects continuing to circle the Earth at increasing distances from each other. But the R-7 had not performed flawlessly, a problem with a fuel regulator just 16 seconds after lift-off forced the engines to consume excessive quantities of fuel and for the orbit to be 500km (310 miles) lower at apogee than intended.

Ironically, when claiming to see the satellite passing overhead as sunlight was reflected from its surface, it was the 26m (85ft) spent core rocket stage that was observed and not the satellite itself, which was much too small to see! To aid in visually tracking it, the core stage had highly reflective panels, but it re-entered the atmosphere on December 2. The satellite's battery ran down on October 26 after PS-1 had completed 326 orbits and it was dragged back down through the atmosphere on January 4, 1958.

Euphoric after a year of intense anguish as to whether the R-7 would ever work as designed, the team under Korolev congratulated themselves. Many had been released

BELOW • For the International Geophysical Year (1957-58), the civilian Vanguard satellite launcher was developed out of the Viking programme. (US Navy)

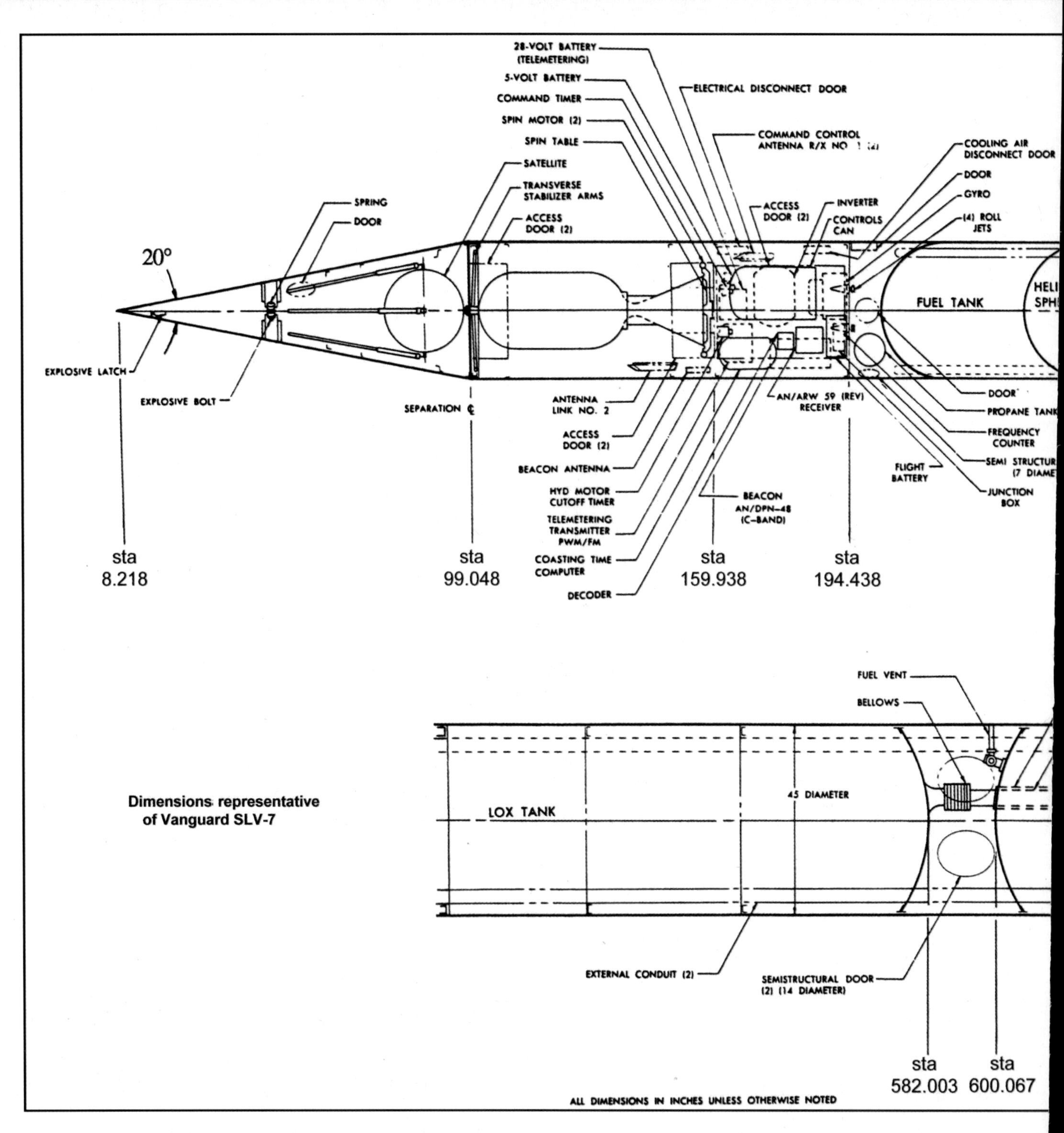

ABOVE • Vanguard was an elegant solution to the problem of developing a satellite launcher and was funded during the 1950s to evaluate new technologies for putting objects in space. (US Navy)

from gulags where they had been incarcerated for minor infringements of the political agenda before working on the national effort. News of the historic achievement was initially low-key, uncertainty as to its success blunting the Soviet propaganda machine until it had been confirmed. But when the international response was clear, laudable acclaim would quickly ramp up for domestic consumption.

The significance of the event was a profound wake-up call to the American public, who regarded Russians as a Slavic people drained by war with Hitler and incapable of highly advanced and sophisticated technologies. It mattered not that they had atomic weapons – hadn't they stolen the secrets from British spies? In Russia it was a matter of intense pride, an effective tool for internal propaganda, an unexpectedly panic-stricken non-communist world suddenly aware that a new super-power had emerged with capabilities only dreamed of in the West.

Outside the select group of Russians who needed to know, the precise size and capabilities of the R-7 were a mystery. Only the weight of the satellite was declared because, as an ICBM the rocket itself was highly classified. What did come as a surprise was the next flight of the R-7 on November 3, when Sputnik 2 was sent into orbit weighing 560kg (1,100lb), inside of which was a dog named Laika in a pressurised compartment. There was no possibility of recovering Laika, and the dog died of heat exhaustion within three days, the details of which only became known many years later.

When it was learned that the dog had died in space, Laika's fate introduced some negative publicity outside the Soviet Union. It stimulated worldwide concern about the use of animals in experiments, prompting organisations fighting for animal welfare to start lobbying against this practice. Some of those still exist 65 years later. Perhaps

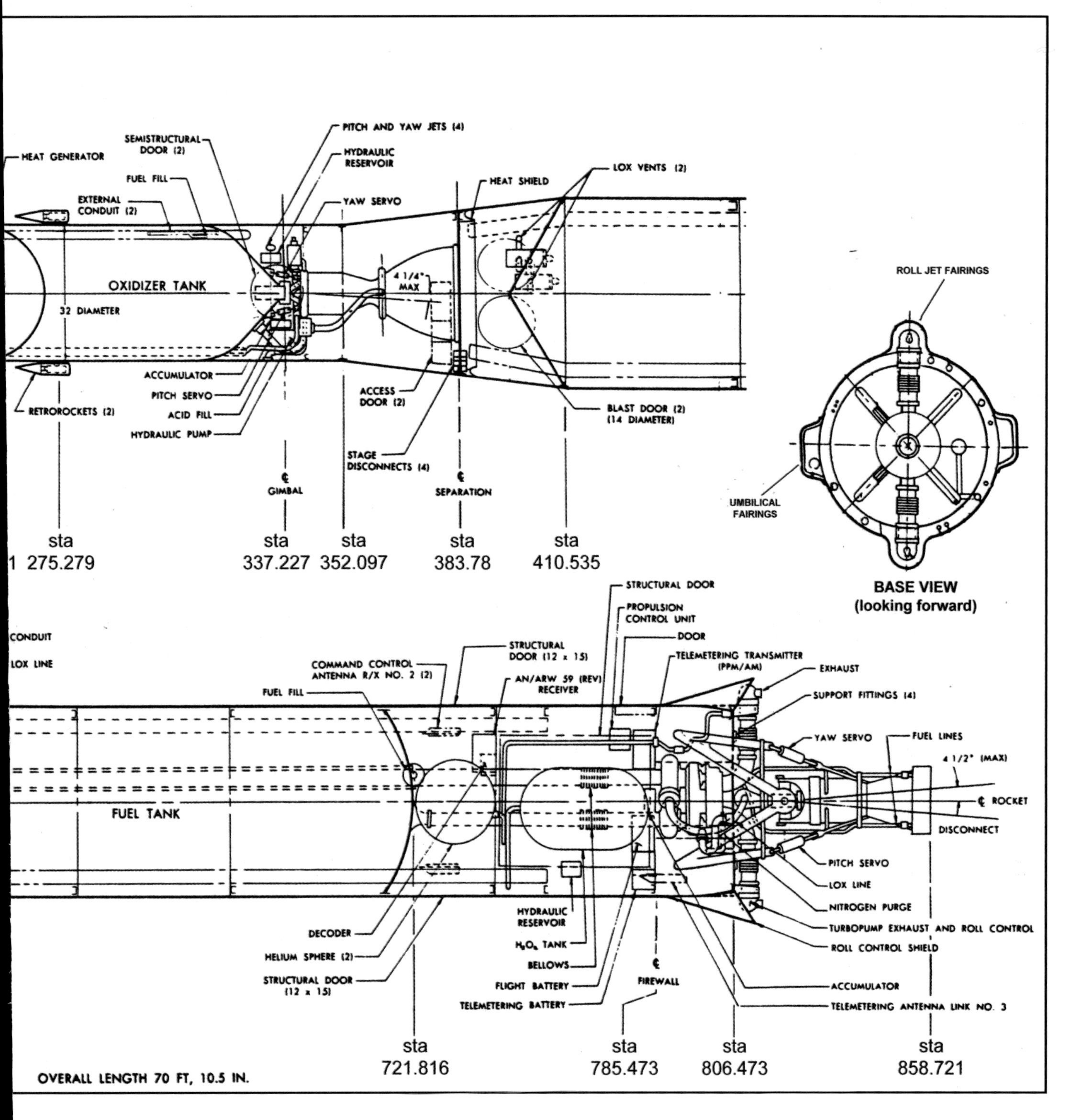

surprisingly, inside the Soviet Union Laika's demise prompted concerns expressed in typical Russian fashion through poetry and music.

Flopnik

Less than three weeks after the launch of Sputnik 1, the third test of the Vanguard rocket took place from Cape Canaveral when TV-2 was flown on a ballistic flight with dummy upper stages. It was to qualify the first stage, to validate pad operations, to further test separation of first and second stages and to provide additional tests with the tracking and telemetry systems. Communications were vital for an object which would be required to send information from science instruments on the satellite. Considerable progress had been made in the development of radio, radar, and tracking systems but the transmission of data (telemetry) was still under development.

TV-2 was planned as a final step prior to the attempted launch of America's satellite for the IGY and it took place on October 23, 1957. The first stage operated as required and the rocket achieved a maximum altitude of 175km (109 miles) and a speed of 6,840kph (4,250mph), the dummy second, third and nose sections returning through the atmosphere and hitting the sea 539km (335 miles) down-range. There had been many delays with the equipment for TV-2 and significant technical issues had prevented the flight occurring for four months after the hardware arrived at the Cape.

Programme managers had wanted several more Vanguard test shots, but pressure was high to get a US satellite into orbit. Thus, it was that TV-3 was prepared for flight with a satellite only 16.3cm (6.4in) in diameter and weighing 1.5kg (3.3lb). Eager as ever to get up close to the action, the national media braved the mosquito-infested swamp and trucked out to Launch Complex 18A (LC-18A),

RIGHT • *Dubbed 'Flopnik' by a cynical media, America's first attempt to launch a satellite on December 6, 1957, ended in disaster. (US Navy)*

where the three-stage rocket waited. Informed that it was to be a repeat of the TV-2 test shot, press personnel knew that this was the attempt to regain a semblance of prestige for the United States.

Ignition of the TV-3 rocket occurred just before 4.45pm local time on December 6, 1957 and for two seconds the rocket moved off the pad. But only to a height of one metre (three feet) before the first stage lost thrust, causing the rocket to collapse back into a ball of fire and thick smoke. To make matters worse, as the nose section and fairing broke away, the little satellite fell off and was consumed in the fireball. To those watching, it was a catastrophe, and, with not a little touch of satire and cynicism, the press soon dubbed it 'Flopnik'.

The US government kept calm and President Eisenhower tried to explain to the American public that there was nothing to worry about. The press had a field day predicting that in celebration of the communist revolution in 1917 the Russians were preparing to detonate an atomic bomb on the Moon. Some news channels panicked and wondered whether the Sputniks were a prelude to a nuclear attack on the United States. Only a very select few in government and contractors at defence institutions sworn to secrecy were aware that the Eisenhower administration had a major spy satellite programme under way. But the Corona programme was at least a couple of years from launch and to the public, it appeared that their government was complacent and in denial.

Delay inevitable

Responding to near panic among the general public, Eisenhower accepted that further delay with Vanguard would be inevitable and that the Russians might attempt something even more dramatic any day. The Army would have to step in with its Project Orbiter proposal. That had been rejected on the basis that it would use the military Redstone rocket, a project masterminded by von Braun at the Army Development Operations Division, Huntsville, Alabama, and which had been developed from the V2. The time had long gone for diplomatic and political niceties. After all, the Russians had used their ICBM to brandish the big stick and Redstone was merely a short-range missile.

For more than 10 years, the German rocket engineers from the V2 programme had been helping the Army develop a range of rockets and missiles, initially based on development of the V2. Redstone was a significant development on that and incorporated technology which the Germans, in the interests of production and making it as simple as they could, had not put on their missile during the war. Developed out of the V2 engine by Rocketdyne, the Redstone motor produced a thrust of 347kN (78,000lb) for two minutes one second and a guidance system for flight from the Ford Instrument Company.

Redstone was capable of throwing a warhead weighing up to 2,860kg (6,305lb) a distance of between 92.5km (57.5 miles) and 323km (201 miles), dependent on the weight of the warhead and the operational requirement. Although classed as a tactical missile by its limited range, it was capable of carrying a very powerful nuclear weapon with a yield of 3.75MT. It had made its first test flight on August 20, 1953.

A second and more powerful missile known as Jupiter was being developed. To test nose cones at higher speeds and greater temperatures for that missile a variant of Redstone was used. It was known as Jupiter-C, for 'composite' because it had a ring of 11 solid propellant rockets on top of the main stage, inside of which were three more of the same type. After separating from the core Redstone stage, the outer ring would fire before separating to leave the inner cluster to accelerate the attached nose cone to greater speed. Von Braun and his team had wanted to place a fourth solid propellant rocket on top of this cluster to get that final stage into orbit.

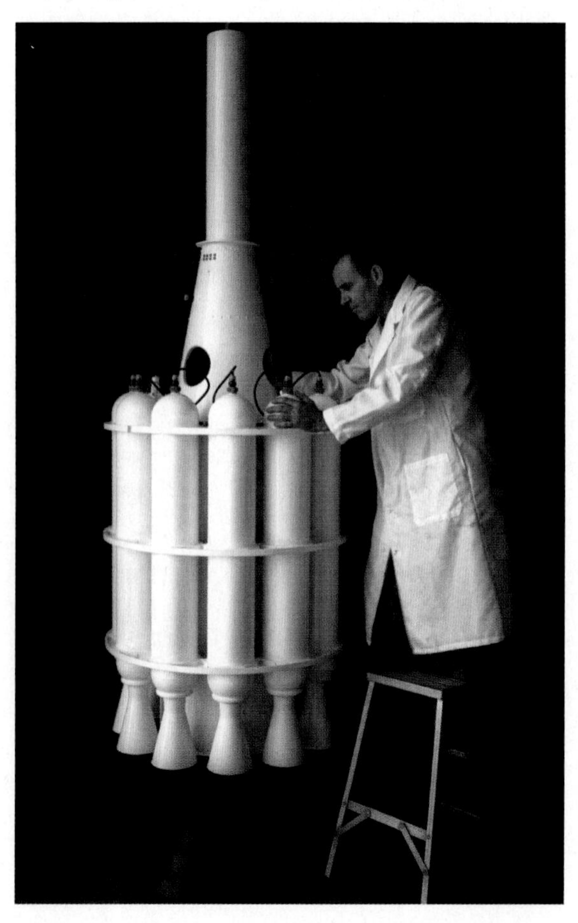

RIGHT • *Von Braun received permission to launch Explorer 1 on a converted Redstone test rocket with three upper stages of solid propellant motors, an instrumented package being attached to the top of the fourth stage. (US Army)*

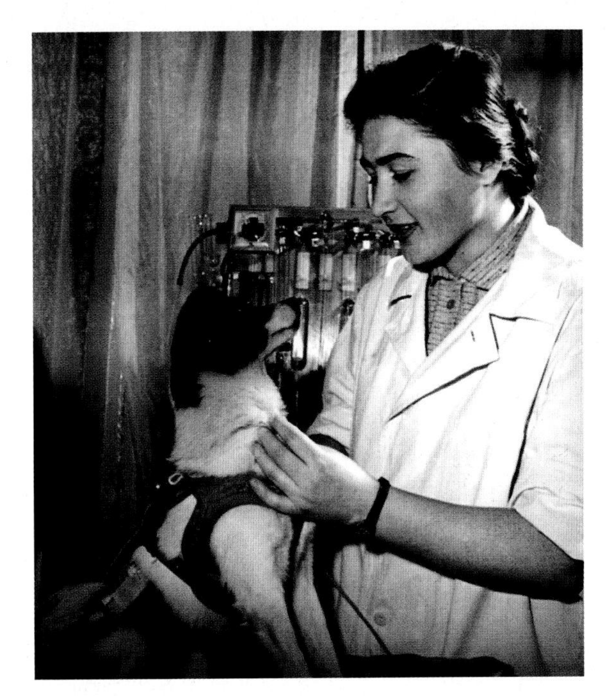

minutes. Contact was maintained until May 23, 1958, when the batteries ran down and it remained in orbit until March 31, 1970, when it decayed back down through the atmosphere and was destroyed.

America finally had a satellite in orbit, but the nation's euphoria was tempered by success with Russia's Sputniks. Neither was US success a foregone conclusion. The second attempt to get a satellite in orbit with a Juno I rocket failed on March 6, 1958, when the fourth stage failed to fire and Explorer II came plummeting back to Earth 3,000km (1,864 miles) from the launch site after the igniter fell out of its socket. Less than two weeks later, in a flight designated TV-4/Vanguard I, the American rocket specifically developed for the IGY placed its first satellite in orbit on March 17, 1958.

As America's second satellite in space, it did little to reaffirm faith in US technology abroad, while the world watched and waited for another Russian spectacle. But Explorer III was up and operating successfully after launch on March 26, a mere nine days after Vanguard I. Another Vanguard attempt failed when the third stage on TV-5 refused to fire after launch on April 29, 1958.

Still the world waited for further news from Russia – and that came less than three weeks later with the flight of Sputnik 3 launched on May 15. This was Object D, which had been planned as the first Soviet satellite but upstaged when precursor and less sophisticated flights ensured that Russia would get into orbit ahead of the Americans.

With a weight of 1,327kg (2,926lb), upon reaching orbit Sputnik 3 was separated from the core stage of the R-7 launch vehicle, where it would remain for almost three years. This was Russia's first science satellite and carried 12 instruments for measuring the near-Earth space environment, a portent of more heavily instrumented satellites to come.

LEFT • Seeking to maximise its propaganda effect, Russia launched a dog named Laika to space in Sputnik 2 on November 3, 1957. (Novosti)

With the failure of TV-3 and uncertainty as to when the next Vanguard could be readied for the attempt, the von Braun team was given approval to rush ahead with its own bid. To further distance it from the overtly military Jupiter and Redstone missile programmes, the four-stage adaptation of the Jupiter-C was rebadged for satellite flights as Juno I, planned adaptations of the Jupiter missile into a potential satellite launcher being known as Juno II. The Army intended to use its rocket engineers to seize the initiative on what was already being perceived as the dawn of the Space Age.

For the Juno I attempt, the fourth stage of the rocket would have an elongated section forward of the rocket motor casing housing science equipment. The two sections would remain connected together, the total length being 204.5cm (80.5 inches) with a diameter of 15cm (six inches) and a combined weight of 13.97kg (30.8lb) of which 8.3kg (18lb) were instruments. The first would be known as Explorer I in a series already being planned by the Army as it competed with the Air Force for management of the nation's space programme.

Data gathering

The science package for Explorer I was managed by Dr James Van Allen of the University of Iowa, and it consisted of a detector for measuring cosmic rays, five temperature sensors and two micro-meteorite detectors for gathering data on impacts from micro-particles. Batteries would provide electrical power and two transmitters would send data to ground stations.

It was the simplest of projects and one which, because of that, had the highest potential for success. Three Jupiter-C Redstone rockets had successfully flown testing nose cones for the Jupiter missile. The last, flown on August 8, 1957, had returned a Jupiter nose cone from space, displayed in the Oval Office at the White House on November 7 alongside Eisenhower beaming in delight.

The launch of Explorer I took place just before 10.48pm local time on January 31, 1958, but 3.48am, February 1 on the Universal Time Clock, which would be standardised for all noted dates for satellite launches. The flight was not without its issues, but Explorer 1 was placed in orbit, circling the Earth with an apogee of 2,550km (1,580 miles) and a perigee of 358km (222 miles) in a period of 114.8

LEFT • Launched on May 15, 1958, Russia's Sputnik 3 was the heaviest satellite launched to date, sending alarm bells ringing. Many Americans feared the size of Soviet rockets. (Novosti)

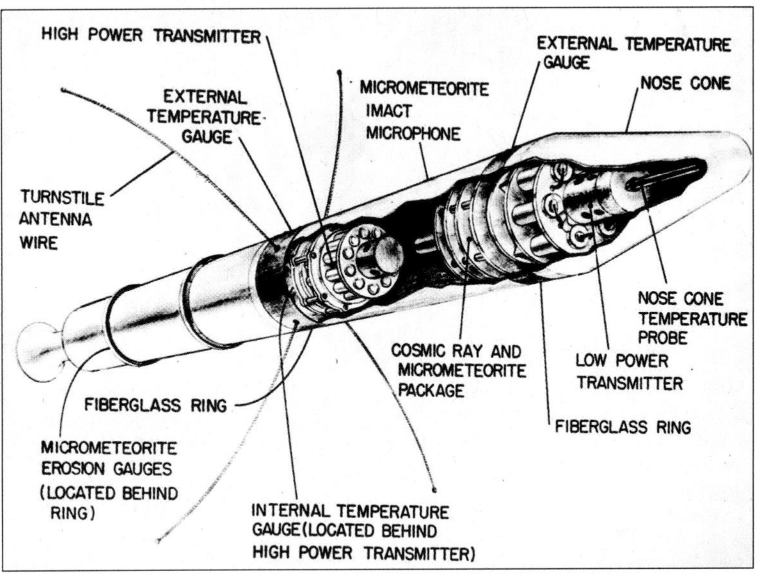

BELOW • The fourth stage of the Explorer 1 launch vehicle as it went into orbit with instruments on the front. (US Army/JPL)

BELOW • Von Braun with his rocket designs. Standing from left to right, Redstone, Jupiter, and Saturn I first stage with upper stages horizontal. (NASA)

MANAGING SPACE

Success and failure go hand in hand in the early days of the space race.

By the first anniversary of Sputnik 1 on October 4, 1958, the Soviet Union had launched three satellites and the United States four. But success had not come easily, at least seven American attempts having failed. In that first year of the Space Age, in America the battle lines had been drawn for fighting out which government body should have control over the nation's space programme. Meanwhile, not for two years after Sputnik 3 would the Russians launch another Earth-orbiting satellite. The reason why would very soon become clear, as the Russians moved inexorably towards a series of flight attempts to the Moon and to the nearer planets – and the first flight of a human in space.

When Russia launched Sputnik 1, it set off a febrile confrontation in the United States between competing groups that had assumed the right to manage the nation's space and missile programmes. Before Sputnik, the Army and the Air Force were in hot competition to carry out this task. In the rush to develop ballistic missiles, each service had its coterie of specialists and experts in science and engineering now driving the new technology. Both services fought for funding from the US Congress, each testifying as to why they should have exclusive control of the new frontier. And both had high talent on tap with more being recruited from the UK and Europe.

Where once they had German General Walter Dornberger as their boss during the war years, in the United States the von Braun group now had General Medaris in charge of their activities, the former Peenemünde engineers working for the US government out of Redstone Arsenal outside Huntsville, Alabama. It was von Braun's Development Operations Division that had produced the Redstone and Jupiter missiles, now developing a series of Saturn super-rockets with a potential thrust greater than that of the Soviet R-7.

Saturn rockets would not be appropriate as nuclear-tipped missiles. They were too big for that and relied on non-storable liquid oxygen with cryogenic upper stages using liquid hydrogen as fuel. Military missiles had to be kept ready for instant launch, which was why the first generation of ICBMs such as Atlas and the R-7 using liquid oxygen as oxidiser would be ineffective until superseded by ballistic missiles using storable propellants. By the end of the 1950s, in the United States the Army clearly envisaged separate branches of rocketry – those for throwing warheads at an enemy and those exclusively for launching satellites.

As mentioned in previous chapters, the Air Force had charge of the Corona spy satellite programme, but the

Army had need of global communications systems for its assumed Cold War role as leader of the free world. To support the widely dispersed bases and facilities from which it operated in many countries, the Army wanted a powerful communications satellite in orbit to relay its communications traffic. Long before Sputnik, communications relay satellites were anticipated as a game changer in connecting various parts of the world by radio and television in a way impossible to achieve through ground transmitters alone.

Concept in hand

When the Department of Defense approved such a system, the Army already had the von Braun Saturn rocket concept in hand as a means of launching such satellites. By the end of 1957, Rocketdyne already had suitable rocket engines on test in addition to research into what would emerge as the world's largest single rocket motor ever built and flown operationally. Designated F-1, it would have a thrust of 6,672kN (1.5million lb) – more than the 20 rocket motors in the five segments of Russia's R-7 combined. The F-1 would be selected for the mighty Saturn V, five of which would power the first stage of the rocket eventually selected to carry astronauts to the Moon.

For their part, the Air Force had an active missile development programme and management of the Corona spy satellites. As the space programme unfolded, Corona was given the publicly declared name of Discoverer, purporting it to be a series of scientific research satellites with a capsule returning biological samples and test instruments from orbit. Nothing about its cameras spying on the Russians and the Chinese. Corona could be launched by a variant of the Thor missile carrying an Agena upper stage, the forward end of which carried the cameras, film spools and a return capsule which would be retrieved in mid-air as it descended.

Meanwhile, the Air Force 'missile czar' was General Bernard Schriever, who fought long and hard against the traditionalists who wanted to retain priority for piloted aircraft. Over time, Schriever would gain control of both the ICBM force and the strategic airborne nuclear deterrent while the Navy got the Submarine Launched Ballistic Missile (SLBM) programme with its Polaris missile and bespoke submarines to support the nuclear triad. But the Air Force would play a major role in developing its Thor, Atlas, and Titan missiles into satellite launchers.

ABOVE • *The Jupiter rocket was developed as a theatre missile but was adapted into the Juno II satellite launcher. (US Army)*

Shuffling the pack

Contrary to public perceptions at the time, when the Russians launched their Sputniks in 1957, the Americans had aces in hand that would propel them to the forefront of the Space Race within a very few years. Unable to share with the general public the scale and breadth of missile

LEFT • *Von Braun was catapulted to fame and national glory by his Explorer 1 flight, which inspired further multi-stage adaptations of the Redstone and Jupiter rockets. (US Army/JPL)*

RIGHT • Gen Bernard Schriever was the driving force behind equipping the US Air Force with ballistic missiles, many of which were adapted into satellite launchers with upper stages, as shown here. (USAF)

RIGHT • All first-generation ballistic missiles were adopted for launching satellites, as with an Able upper stage here being attached to a Thor rocket in 1958. (USAF)

BELOW • The Thor missile, here being operated by RAF crews prior to its deployment to the UK, formed the basis for satellite launchers with upper stages, eventually named Delta. (USAF)

and satellite programmes underway, the Eisenhower administration went down as a lacklustre government in denial over the perceived threat. And there was much political opportunism to be gained as the opposition fought to deny the government credibility, which would endure right up to the presidential election campaign launched in 1960.

Eisenhower insisted that civilian and military space programmes should be distinctly separate and be seen as such. Moreover, he distrusted the armed services in controlling the policy directives that each sought to empower with government funding and embolden with classified programmes. Accordingly, on February 7, 1958, he formally established the Advanced Research Projects Agency (ARPA) within the Department of Defense with a mandate to push ahead in space, defend the nation from Soviet attack and support technologies that 'fall between the cracks'. It was to bring together all the separate strands of military and non-military research and development projects so that a balanced and proportionate use of national resources could proceed for the national good, rather than the vested interests of each armed service.

As the furore over Russian space successes showed little sign of dissipating, however, the US Congress began a series of hearings to decide on the future organisation of the various space concepts. Politicians hounded by a concerned and agitated electorate intensified debate and discussed with a seemingly endless torrent of experts just what to do about the propaganda coup. Evidence about the present state of the missile, rocket and nascent space industry was sought from far and wide. The great and the good lined up in droves to give testimony, each urging upon the political establishment urgent action to counter the tide of Soviet successes.

A leading advocate of a fresh approach to management of space activities, Senator Lyndon B Johnson, had begun hearings into the challenge as early as November 25, 1957 and on February 6, 1958 a special Senate committee was set up chaired by Johnson. With guidance from a range of specialists and policy makers, on April 2 Eisenhower sent draft legislation to Congress for a National Aeronautics and Space Agency to manage non-military programmes.

In drafting the new law, Eilene Galloway from the Congressional Research Service thought it would have more power if it became an 'Administration' rather than an 'Agency' under a director.

On July 29, 1958, President Eisenhower signed into law the National Aeronautics and Space Act, establishing the National Aeronautics and Space Administration (NASA) out of the National Advisory Committee for Aeronautics (NACA) which had been formed in 1915. From October 1, 1958, the several field centres of the NACA would now become NASA centres with distribution of new roles and responsibilities growing out of the existing functions at each location.

For much of its life, the NACA had been committed to research on aerodynamics, aircraft design and engineering and to associated scientific and technological capabilities. Since the Second World War, its interest in rocket propulsion and high-speed flight had led it to assume increasing responsibilities for the X-series research aircraft which had seen the first flight through Mach 1 in level flight by Major 'Chuck' Yeager in 1947. Its most current effort on that frontier was the hypersonic X-15 programme, which would take manned flight beyond Mach 6. Various NACA field centres had also conducted research into the potential for manned flight.

Both the Army and the Air Force had their own concepts for placing humans in orbit and these would converge in work conducted by the NACA and inherited by NASA. The main activity was at the Langley Aeronautical Laboratory, renamed the Langley Research Center where the work evolved into Project Mercury as NASA took over the programme that would place the first American astronauts in space. Langley was the oldest of the NACA facilities, along with the Ames Aeronautical Laboratory (renamed Ames Research Center) and the Lewis Flight Research Laboratory (now the John H Glenn Research Center).

Facilities absorbed

Other facilities were absorbed into NASA as its responsibilities grew, most notably the High Speed Flight Research Station at Edwards Air Force Base, California, renamed the Dryden Flight Research Center and now known as the Armstrong Flight Research Center. During 1958, the Jet Propulsion Laboratory was signed over to NASA in a shift from rocket motor design and development to planetary research, a task in which it soon became the global leader. The following year, NASA set up the Goddard Space Flight Center, a completely new facility dedicated to work on science and applications satellites and for the development of related space vehicles.

The job of developing and producing the massive Saturn rockets took the Development Operations Division at the Army's Redstone Arsenal, Huntsville, Alabama, to NASA in 1960 when it was renamed the Marshall Space Flight Center with von Braun appointed director. To manufacture Saturn stages and to provide facilities for the contractors employed to do that, NASA took over the Michoud Assembly Facility and in 1961 announced development of the Mississippi Test Facility, now renamed the John C Stennis Space Center, for testing the rocket engines.

Dedicated site

While launch facilities at Cape Canaveral were retained by the Air Force when President Kennedy made the decision in 1961 to mobilise the Apollo programme and send astronauts to the Moon, a dedicated site at that facility was built and named the Kennedy Space Center after the death of John F Kennedy in November 1963. Apollo also stimulated plans for a dedicated Manned Spacecraft Center built just outside Houston, Texas, and operational from 1965.

Over time, NASA grew out of a small organisation employing 8,420 people to become America's hub for space exploration and applications and the civilian control and leadership role that it was set up to achieve, becoming a world leader in aerospace technology and mission management. At its height during Apollo, it employed more than 30,000 people and its programmes supported a contractor workforce of more than 400,000 throughout industry.

Paramount to its primary role has been the open and accessible dissemination of information and results, feeding the aerospace industry of America with enabling data to mobilise national resources. In doing that, it provides technical notes to manufacturers, affords open access to its information, and provides advice to engineers, scientists, and technologists both inside and outside the space programme. At its inception, the Act that brought it into being required NASA to achieve and then to maintain national leadership in space technology and that remains its core role today, but it has also inspired similar aspirations far beyond the borders of the United States.

ABOVE • *Former President Harry Truman (left) enthuses with NASA administrator James Webb over the prolific array of US satellite launchers. From right: Scout, Juno II, two Thor derivatives, four Atlas/upper stage launchers, two Saturn I and a massive Nova, the latter never developed. (NASA)*

BELOW • *Eclipsed at first by Soviet rockets, US manufacturers soon acquired bigger and more powerful motors such as the F-1, five of which were used in the first stage of the Saturn V, which would launch men to the Moon. (Author's collection)*

ABOVE LEFT AND ABOVE RIGHT • *Founded in 1915 to boost America's aviation industry, the National Advisory Committee for Aeronautics (NACA) was re-established as NASA effective from October 1, 1958. (NASA)*

With interest in satellites growing fast after the initial launches by Russia and America, a group of European scientists came together to set up the European Space Research Organisation (ESRO), which came into existence in 1964 with nine continental nations and the UK as signatories. The purpose of ESRO was to develop a unified plan for launching sounding-rockets and to develop science satellites which would be sent into orbit on Europa rockets. It also served as the focus for ground-based research and for developing an exchange of information and international research.

It turned out that satellites developed by ESRO were all launched on American rockets under a reimbursable agreement whereby NASA managed a deal between the European consortium and the launch providers. These were usually Thor or Thor-Delta rockets provided by the Douglas Aircraft Company. But the future opportunity for ELDO to provide independent launch services for European satellites collapsed when in 1966 the British withdrew Blue Streak as the first stage for the Europa rocket. The performance of Europa had always been questionable when comparing development cost with the potential in its capabilities but this single decision by the UK changed the future for European space endeavours forever.

But the very shape of space activities changed dramatically, in that the type of orbit required for most satellites did not match the capabilities of Europa, Nevertheless, without a first stage, the concept had no future. Initially funded 40 per cent by the British, this was reduced to 27 per cent in an attempt to encourage sustained commitment, but support was still withdrawn, leaving a few rockets funded to fly test launches. They were initially flown from Woomera in Australia before moving to Kourou in French Guiana, from where the Europa satellite launchers were to have been sent into space.

Eleven flights were attempted between June 1964 and November 1971 and while the Blue Streak performed well, the upper stages proved troublesome and the last six were failures, the final attempt being from Kourou. The demise of Europa left a bitter taste in affairs between the political leaders in the UK and European countries so heavily invested in an independent satellite launch capability. Largely out of a determination that they, unlike the British, refused to rely on the Americans for launching their satellites, the French organised a new launcher programme.

Accepting the limitations of ESRO and the inability of ELDO to provide a launch vehicle, both organisations were merged into a new European Space Agency (ESA), which was set up in 1975. France organised a new satellite launcher in the form of Ariane, which was one of two flagship programmes for the new body, the other being Spacelab from Germany, a manned laboratory designed to fit inside the Space Shuttle Orbiter and support scientific research prior to a future, independent space station.

Thus, was formed the structure within Europe which today supports a vibrant organisation boasting 22 full members, of which the greatest contributors are France (24.5 per cent), Germany (21.1 per cent), Italy (14.1 per cent) and the UK (9.1 per cent). Members contribute sums approved by a Council of Ministers and each country gets back in work a percentage of contracts proportionate to their financial contribution. This allows member states to decide how much they want to get out for their own industries and research institutions by determining what they can afford to put in.

A European challenge

With the Cold War at its height, in the late 1950s the United States and the UK entered into an agreement whereby American rocket companies would provide Britain with technology for a ballistic missile known as Blue Streak. Designed to reach the Soviet Union from bases in the UK, as a missile the rocket was cancelled in 1960 but offered to a European group of nations under the European Launcher Development Organisation (ELDO) founded in 1962. Blue Streak was to provide the first stage of a multi-stage satellite launcher named Europa, with the second stage provided by France and the third stage by Germany.

Although recovering from a war that left most continental European countries in ruins, there was considerable talent in aerospace technologies and the outstanding strides taken by Germany in missile and rocket technology fed a post-war desire for national reconstruction around those assets. Accordingly, in France and in the UK, there was enthusiasm for those technologies. But not among many politicians who failed to recognise the way the future of science and engineering was changing direction, with jet aircraft, supersonic flight, guided missiles, and long-range rockets the order of the day. It was left to industry and leading scientists to educate politicians reluctant to look afresh at how the post-war world would develop, their words frequently falling on deaf ears.

BELOW • *The emerging space industry was built on the capabilities of the US aviation industry with specialised services developed for satellites and launch vehicles. (NASA)*

Diversity in space

Between 1957 and 1970, 11 countries launched satellites, five of which (Russia, the US, France, Japan, and China) were sent into space on their own rockets and six launched by the United States on a reimbursable basis. Seeing a marketing advantage in the increasing number of different rockets available, the US offered to launch civilian satellites for other countries on the condition that none of them would compete commercially with US satellites or the activities of domestic companies. This only became relevant in the 1970s when communications and broadcast satellites were launched by telecommunications companies, the US refusing to launch foreign satellites which would take revenue from the US operators.

The UK was the third country to fund a satellite when it had the US launch Ariel 1 on a US Thor-Delta rocket on April 26, 1962, from Cape Canaveral. It was the success of this that consolidated the political decision in 1964 to develop a national launch vehicle named Black Arrow, capable of placing a satellite weighing 144kg (317lb) in a low Earth orbit. Several governments came and went, and the project was delayed, but only one out of four attempts from June 28, 1969 was successful. On the final attempt, the Prospero satellite was sent into orbit on October 28, 1971, from the Woomera test range in Australia.

It very nearly didn't happen. Just three months earlier, the programme had been cancelled by the British government but as the stages had been shipped, the attempt was allowed to go ahead, leaving the UK to be the only country to have successfully launched its own satellite on a home-built rocket and then cancelled further development. Along with cancellation of Britain's contribution to ELDO and withdrawal of the Blue Streak as first stage for Europa, it placed a cloud over further involvement by the UK in space activity until the country folded its national activity into the European Space Agency.

The British achievement was late in the day, however, as France eclipsed anything under way in the UK. Built on the back of President Charles de Gaulle's commitment to give France an independent nuclear deterrent, from the late 1950s the country's best scientists and engineers developed rockets, ballistic missiles, and the infrastructure for an emerging space programme. The political commitment was total, and de Gaulle made it a national goal, lifting France from the ravages of war to a new and technologically advanced state with aerospace at the forefront.

ABOVE • *A wide range of new technologies were developed for launches and for satellite operations, as seen here with the launch blockhouse for an Atlas-Agena flight from Cape Canaveral. (NASA)*

BELOW • *Within a few years, the proliferation of launch sites for ballistic missiles on test and satellites launched into space from Cape Canaveral transformed a mosquito-ridden swamp into a major industrial site. (NASA)*

RIGHT • The UK built the Blue Streak ballistic missile at the Stevenage facility of the then Hawker Siddeley Dynamics. (HSD)

At the heart of that was a desire to distance the country from an unalloyed connection to the United States that de Gaulle perceived was inhibiting the rest of Europe from carving a separate identity in the post-war world. During the 1960s, France would establish a fully independent nuclear deterrent, basing missiles in silos on land, in submarines at sea and aboard aircraft in the air, while withdrawing France from the military wing of NATO. It was this stoic determination to restore national pride, demonstrate European leadership and maintain independence in science and technology that gave France a national launch vehicle programme named Diamant.

Not unlike the American Viking and Vanguard launcher programmes, Diamant grew out of a prolific range of sounding-rocket designs which gave France both the knowledge and the experience to build and launch its own satellites. It helped that France had its own programme of missiles for the national nuclear deterrent as a lot of common technology existed between the two requirements and this cut development time for both. It also helped that the government invested heavily in national research and development facilities, the key to a successful, fast, and economical space programme.

BELOW • The Blue Streak missile was cancelled and repurposed into the first stage of the Europa launch vehicle (left). (Author's collection)

Three versions

France's national space agency, the Centre National d'Études Spatiales (CNES) developed three versions of the Diamant satellite launcher. CNES had been formed on December 19, 1961 and would be responsible for all space activity in France, managing sounding-rocket and satellite launches from a base at Hammaguir in French Algeria until a move to French Guiana and the Kourou site, where it remains today for flights with Ariane and Soyuz rockets.

Diamant A was flown four times from November 26, 1965 to February 15, 1967 and achieved success on the first launch when it orbited the Asterix satellite weighing 42kg (92.6lb). Three more satellites followed, all orbited successfully, although the third had a lower orbit than planned. The improved Diamant B flew from Kourou on five launches between March 10, 1970 and May 21, 1973, three of which were successful in orbiting satellites. These were followed by three Diamant BP4 launchers from Kourou

with greater performance and which successfully orbited satellites between February 6 and September 27, 1975.

Following the cancellation of Europa and the establishment of the European Space Agency, national interests folded into this pan-European organisation and continental countries pooled resources to develop the Ariane satellite launcher. Based on the highly successful Diamant series, in which only two out of 12 flights were failures, France was to lead in developing what would quickly become one of the most successful satellite launchers yet developed.

Not only did it capitalise on French expertise and success in aerospace programmes across a wide spectrum, but it also stimulated the formation of Arianespace SA, the world's first commercial launch provider, marketing the services of the Ariane series of rockets for an increasingly wide range of customers around the world. Ariane 1 flew successfully for the first time on December 24, 1979, followed by five successive generations preceding the Ariane 5, which made its first successful flight on October 30, 1997.

Ariane 1 could put 4,850kg (10,690lb) in low Earth orbit but Ariane V had the capacity lift 20,000kg (44,000lb) into the same path. But the later Ariane launchers had evolved to optimise their performance for lifting satellites toward a geosynchronous orbit approximately 33,400km (22,300 miles) above Earth. At that distance, satellites orbit the Earth in 24 hours and therefore appear to remain stationary above a single spot. In all, Ariane launchers have made 260 flights since 1979, of which 116 have been made with Ariane 5 achieving a success record of more than 95 per cent. Ariane 6 is expected to replace it in early 2024.

Asia's space race

In its drive to develop a post-war economy, when the Second World War ended in 1945 Japan sought to capitalise on its industrial and scientific capabilities by developing a consumer economy with motor vehicles and motorcycles as well as ships and civil aircraft. It also engaged in academic research using sounding-rockets for a study of the Earth's atmosphere and for participating in global research projects. The first formal activities began at the Institute of Industrial Science, University of Tokyo, with Pencil and Baby sounding-rockets tested in Akubunji, Tokyo and at the Akita Rocket Testing Center in Michikawa Beach.

From these modest beginnings, research into sounding-rockets and the exploration

LEFT • Models of the UK Black Arrow rocket and Prospero satellite on display at the London Science Museum. (Dtodd1/ Wikimedia commons)

BELOW • A Europa test launch from the Woomera, Australia, site before flight operations shifted to Kourou, French Guiana on the coast of South America. (HSD)

BELOW • The Woomera test site in Australia was developed into the launch site for Blue Streak test flights and for early launches of the Europa rocket. (Author's collection)

ABOVE • *France's Diamant satellite provided the basis for the Ariane launch vehicle which became one of the two major flagship initiatives of the European Space Agency. (Author's collection)*

BELOW • *Development of the Ariane put Europe on the map of space launch providers, securing a niche through five generations of rocket, from the Ariane 1 capable of sending 1,850kg (4,080lb) to a geosynchronous transfer orbit to Ariane 5 and a capacity for sending 10,865kg (23,953lb) to the same trajectory. (Arianespace)*

of near-Earth space followed, with more advanced solid-propellant rockets such as Kappa, which played a leading role in global measurements of the outer atmosphere. There were several versions of Kappa, some carrying packages of science instruments weighing up to 60kg (110lb) to heights of more than 740km (460 miles).

In February 1962, the Kagoshima Space Center was set up to support more advanced flights and in 1964 the Institute of Space and Astronautical Science (ISAS) was formed in the University of Tokyo with aspirations to develop bigger rockets and launch satellites into orbit. The Lambda sounding-rocket provided the development step for the Mu-series, which would result in the launch of Japan's first satellite on February 11, 1970. Named Ohsumi, it was a technology demonstrator making Japan the fourth country after the Soviet Union, the United States and France to build and develop its own satellite and launch capability.

Amid a growing awareness that space technology was providing an expanding base for science and technology, on October 1, 1969, the Japanese government set up the National Space Development Agency (NASDA) to integrate several potential satellite projects for the peaceful exploration of space. At the time, Japanese law prevented the development of military technology which had a potential for offensive operations, hence the emphasis on 'peaceful' activity as enshrined by legal statute.

NASDA immediately began development of the N-series satellite launchers, which would be flown from a new complex on the southern island of Tanegashima, arguably the most beautiful launch facility in the world, usually accessed by air from the mainland. From the N-series came the H-series, making Japan a global leader in space technology in satellites and in space probes to the Moon and the planets.

During the 1980s, Japan supported human space flight through its participation in the International Space Station managed by NASA, along with Russia and the European Space Agency, all of which have supplied modules to the station and sent astronauts to the orbiting laboratory. As noted in the following chapters, Japan has made major strides in developing space technology for useful applications benefitting the economy, people's lives, the environment and giving a better understanding of the planet's weather systems.

In October 2003, the government integrated ISAS, NASDA and the National Aerospace Laboratory into the Japan Aerospace Exploration Agency (JAXA). To date, Japan has launched almost 100 satellites into orbit and has been a major contributor to international space endeavours.

The fifth country to launch its own satellite, China began its interest in rocketry and space research during the 1950s in the controversial reign of Chairman Mao Zedong. With help from the Soviet Union, the newly formed communist state followed Russia in a programme of major industrialisation, replacing centuries of agrarianism with a push towards urbanisation and a new technological revolution. The space programme that emerged in the 1960s was built around the purchase of Soviet missiles as the country strove to become a nuclear power – for energy production and weapons, with rockets from Russia and scientists and engineers returning from the United States.

'We too need satellites'

The real stimulus came immediately after the flight of Sputnik 1, when Chairman Mao declared that 'we too need satellites', organising Project 581 to get a satellite in orbit in 1959 for the 10th anniversary of the founding of the country's communist rule. That proved too ambitious. Russia had used its big R-7 ICBM to send up its Sputniks, but China had nothing like that and diverted the effort into a major sounding-rocket programme as a preliminary step. Not before early 1960 did the T-7M begin probing the upper atmosphere while an ideological split severed China from access to Russia's advanced rocket and space technology transfer.

China connected its ballistic missiles programme with its aspiration to become a space power and as that technology evolved, the satellite launcher came along as an addition. On April 24, 1970, the 173kg (381lb) Dong Fang Hong satellite was placed in orbit and the nation joined the space club. Not to be outclassed by Russia or its ideological competitors in the West, in 1967 China decided to begin a manned spacecraft project with the objective of sending humans into orbit. The ambition was outclassed by its capability and not until October 15, 2003, was the first Chinese astronaut sent into space, 41 years after the Russians and the Americans. It was the start of what has evolved into a fully manned space station and plans to put humans on the Moon by 2028.

Between the first satellite and the first Chinese astronaut, the country developed a range of medium and

heavy-lift launch vehicles in parallel to a burgeoning development programme for military rockets and missiles. Its space programme, like that of the former Soviet Union, is enmeshed within the military and national security interests of the country and it is difficult to find a dividing line between civil and military alliances and responsibilities. Supporting it all, however, is a wide range of institutions and technology facilities supporting the several separate launch sites.

Overall management and control of the national space effort is the China National Space Administration (CNSA) formed in 1993, ostensibly for 'civilian' space activity and for international co-operation. Unlike its American equivalent NASA, the CNSA does not conduct space operations, that being under the management of the China Aerospace Science and Technology Corporation.

The overall objective of China's space programme is to support the political aims of the national government, to serve as a polarising influence on the direction of the country's aerospace technology and to project the aspirations of the nation's leadership through influence and with joint ventures. In that, it is akin to both the former Soviet Union and to NASA in the United States. By mid-2023, China had launched more than 540 satellites for its own domestic programme and foreign operators.

On July 18, 1980, India became the sixth country to build and launch its own satellite. Named Rohini RS-1, it weighed 35kg (77lb) and was placed in orbit by the four-stage, solid propellant Satellite Launch Vehicle (SLV). This satellite was the first in a series of four which provided data on the performance of the terminal stage and helped define a technology development programme which was both systematic and timely.

India had been conducting research into rocket propulsion since achieving independence from Britain and put considerable resources into the development of key technologies. With only one totally successful flight, the SLV was a painful learning curve, as was its successor the Augmented Satellite Launch Vehicle (ASLV), of which four were launched between 1987 and 1994, with again only one success in four attempts. The more capable Polar Satellite Launch Vehicle (PSLV), with a payload capacity of up to 1,400kg (2,900lb) to a polar orbit followed, with a first flight in September 1993. By mid-2023, it had achieved 54 successful flights out of 57 attempts.

Stationary appearance

Next for India was the Geosynchronous Satellite Launch Vehicle (GSLV), specifically designed to place payloads of up to 2,500kg (5,500lb) into a geosynchronous transfer path, a point where the satellite could fire a small rocket motor and circularise its orbit so that it appears to remain stationary over one spot on Earth. It can also lift 3,000kg (6,600lb) into low Earth orbit. A much more complex launch system using large strap-on solid propellant boosters and a cryogenic upper stage, the GSLV is a challenging programme. First launched in April 2010, by mid-2023 it had successfully launched eight times out of 14 launches.

The long road to success for India's space programme began around a century ago when ground-based radar was used to probe the ionosphere for measurements related to radio communications. That matured into a national resolve to invest in modern technology after the country gained independence in 1947, gaining traction with the launch of Sputnik 1 in October 1957. In no other country was the Sputnik effect more evident than India, a country keen to maintain a non-aligned geopolitical stance balancing trade and co-operation with both

BELOW LEFT • *China has invested heavily in its space programme. Here, a Long March 2F launches a Shenzhou manned space vehicle. (ForrestXYC/Wikimedia commons)*

BELOW • *Japan has a vibrant and productive space industry, with launch sites on the southern island of Tanegashima, arguably the most beautiful complex of any space-faring country. (NASA/Bill Ingalls)*

the communist world of Russian states and the Western democracies.

Prime minister Jawaharial Nehru set up the Indian National Committee for Space Research in 1962 but what it lacked in centralised co-ordination it made up for in the collective contributions of related institutions and factories producing small rockets. Flights with early Indian sounding-rockets were launched from the Thumba Equatorial Rocket Launching Station near the southern tip of the country and considerable experience was acquired through the development of these devices.

The successful Moon landing by Apollo 11 in July 1969 put astronauts Neil Armstrong, Buzz Aldrin and Mike Collins on a world tour which greatly influenced many countries to get involved in space research, re-charging the precedent set by Sputnik 1. The landing played no small part in stimulating the government into forming the Indian Space Research Organisation (ISRO) on August 15, 1969, two months before the crew visited the country. The nation had erupted with 'Apollo fever' and one million people thronging the streets of then Bombay (Mumbai since 1995) enthralling many young people who would later work for ISRO, or some other organisation related to the nation's space programme.

Three years later, an important administrative step was taken with the establishment of the Department of Space within central government, and this aided a co-ordinated and efficient structure for an increasingly ambitious strategy. Throughout the last 50 years, India's programme has focused on remote sensing of the Earth and its environment, space science, satellite communications and a modest programme for exploring the Moon and the planets.

Russian roulette

Following its initial achievements with Sputnik's 1-3, Russia moved quickly to capitalise on its success and developed space probes to the Moon, to Venus and to Mars. It also sought to put the first human into orbit, and it achieved that on April 12, 1961 with the launch of Yuri Gagarin. This stimulated renewed energy in the US space programme and put NASA on course for the first manned landing on the Moon, which it would achieve in July 1969.

Perhaps surprisingly given the publicity awarded to Russian achievements, the response of the Americans was both bolder and more ambitious than anticipated. Reacting accordingly, in several areas it was the Americans that pushed the Soviet space programme in the direction it took, with a competitive attempt to get its cosmonauts to the Moon before NASA's astronauts and to upstage US plans for both military and civilian space stations. The US would cancel its manned military station, but the Russians rushed ahead on both fronts.

Progress at NASA with the Apollo Moon goal shocked the Russians, who committed to a race to the lunar surface as late as August 1964 and only after excessive deliberation. Counter-intuitively, the top-down structure of the Soviet government presided over several separate design bureaus, each competing with the others for favours and priority funding. At the core was Korolev's OKB-1 design bureau fighting off opposing design teams, which only served to stultify the overall strategic objectives expressed by the political leadership.

With individuals supporting different teams, progress was sluggish and counter-productive and when Korolev died in January 1966, the effort fragmented further, separate design bureaus fighting for prime favours and lucrative contracts. The Russians were unsuccessful in developing their equivalent to NASA's mighty Saturn V, abandoning their manned Moon landing plan in favour of Salyut and Mir space stations. In other objectives, Russian scientists achieved outstanding success in the exploration of Venus but very little in their missions to Mars.

By the end of the Soviet system in 1991, the Russian space industry and the financing it required to stay current was running down. The United States offered partnership in the International Space Station (ISS), to which Russia will remain a participant until it withdraws by the end of this decade. As NASA transitioned from the Space Shuttle to commercial vehicles from SpaceX and Boeing for accessing the ISS, Russian Soyuz spacecraft kept the traffic lanes

BELOW • China is developing a wide range of launch vehicles for an expanding space programme with broad ambitions and is developing a variety of rocket types, its Long March series displayed here at an industry show in 2021. (Kirill Borisenko)

LEFT • India's space programme is based largely around applications and benefits to its people, including development of the GSLV launch vehicle capable of placing 2,500kg (5,500lb) satellites to a geosynchronous transfer path for environmental, resource and atmospheric studies. (ISRO)

flowing for astronauts and cosmonauts visiting the 400-tonne orbiting laboratory.

After the collapse of the Soviet Union, management of the national space programme transitioned from the Ministry of General Machine Building where it had been located since 1955 to the Russian Space Agency on February 25, 1992. By the end of the decade, it had been renamed the Russian Aviation and Space Agency and from 2004 it was known as Roscosmos, or Federal Space Agency. Co-operative ventures with Western countries collapsed after Russia invaded the Ukraine on February 24, 2022, with many projects and programmes cancelled as a result.

BELOW • The rockets used to launch early Russian satellites were developed along the line of R-7 ballistic missiles, as seen here, but also diversified into a wider range of different launch vehicles. (NASA/Peter Gorin/Emmanuel Dissais)

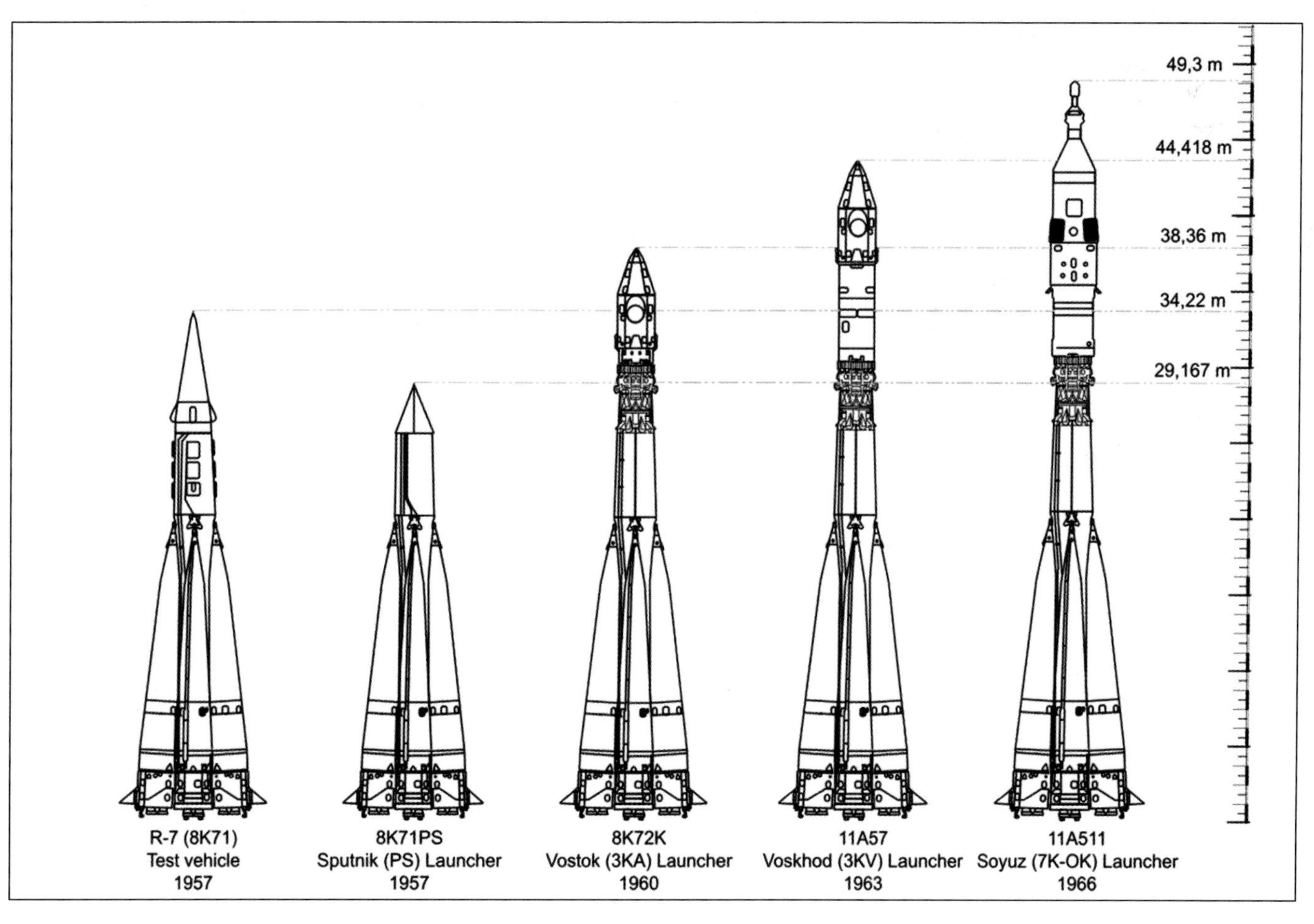

SPIES IN SPACE

...and scientific research go hand in hand. ...ervation from on high has been around for a long time.

ABOVE • *Aircraft such as this featherweight Convair B-36D carried out frequent covert spying operations over the Soviet Union during the 1950s. (USAF)*

As noted previously, the United States had an active, albeit covert, military satellite programme code-named Corona, underway long before Sputnik 1, a contemporary endeavour to its open and widely publicised Vanguard science satellite programme. The Russians had nothing of this sort, and it would be several years before they did. The price paid was that to the general public America was way behind when in reality it had a menu of intensively studied space concepts for optical, radar-based and signals-intelligence (SIGINT) projects. Some of those were funded and a few were in advanced stages of development when the Russians launched Sputnik 1.

The need for detailed intelligence about the true level of technological capabilities possessed by the Russians was more pressing than ever. The first operational flight of a U-2 spy plane had taken place on June 20, 1956 and the Soviet rocket test site at Baikonur had been photographed in August 1957, even the launch pad where Sputnik 1 would fly from. The Russians designated the site as Tashkent-50 but to the CIA it was Tyuratam, after the nearest railway station. Soon it would fall to an RAF pilot to fly the U-2 from RAF Watton in East Anglia and the Russian site would be immortalised as Baikonur.

The most opportunistic flights originated at Peshawar in Pakistan, taking them up to the Sverdlovsk region before diverting to bases in Iran or other locations in the Middle East. Other flights being planned would take place right across the central USSR from Peshawar to Bodo in Norway, a distance of 6,500km (4,040 miles) and a flight time of more than nine hours. Soviet air defences were getting better, and the international consequences of a downed U-2 were incalculable. There was a new urgency with each overflight.

ABOVE LEFT • *Heading the USAF Strategic Air Command, Gen Curtis LeMay argued widely for expanded spying operations, which encouraged the development of satellites to carry out this activity. (USAF)*

ABOVE RIGHT • *President Eisenhower approved the top-secret development of spy satellites in the Corona programme, publicly named Discoverer. (White House)*

Increasingly nervous about using them for these vital photo-runs, Eisenhower grounded the U-2 from October 13, 1957, to March 2, 1958, resumption of which with an overflight across the Far Eastern regions of the USSR brought a formal protest from Khrushchev. Flights were again halted on March 7, and it would be 18 months before they resumed. Following a limited series of spy flights over Russia, a U-2 piloted by Francis Gary Powers was shot down near Sverdlovsk on May 1, 1960.

The ironic twist was that on August 19, 1960, the very day the US obtained the first spy satellite pictures of Russia, in Moscow Gary Powers was given a ten-year prison sentence for spying from the air in Soviet airspace. The Corona programme had taken a long and tortuous course in the six years of development it took to build a credible reconnaissance and surveillance satellite system, diversifying into a wide range of space-based intelligence gathering, during which it would transform the way national intelligence was obtained and how it was used.

Defining intelligence

When von Braun had first been interrogated in May 1945, he laid out a sequence of developments which would become possible given sufficient material and political support. The information he disclosed about possibilities for satellites and space vehicles fired the imagination of personnel at the US Navy. On March 7, 1946, it proposed a joint Navy-Army programme to develop a satellite for reconnaissance and surveillance. The Navy had for long had a deep commitment to intelligence gathering. The offer of co-operation with the Army was a way of circumventing the severe budget cuts following demobilisation.

In early 1949, a classified conference studied the psychological impact such a programme might have on the USSR. The satellite was viewed by some as a tool for unseating the Soviet regime by periodically disclosing information thus gained about the country to destabilise public enthusiasm for the closed system. If co-ordinated with the foreign broadcast services through the Voice of America, drip-feeding a disillusionment among the Russian people, it could bring about recrimination and the possible collapse of their political system.

As satellites became plausible, the Air Force and the Central Intelligence Agency (CIA) established criteria for producing a set of requirements that could define a specification detailing technical standards for optical systems or radar. Developments with aerial reconnaissance cameras had been comprehensive during the war and there was an opportunity to apply that technology to satellite systems. The generally agreed criteria defined an increasingly sophisticated set of separate strands.

Several levels of intelligence called for sequentially increased levels of optical resolution. These included detection, general identification, precise identification, description, and technical intelligence. The degree of resolution required to fill each of these levels differed according to the target. Detection required determination of a general type or class of object, while general identification required the analyst to be able to decide on the type of target observed.

ABOVE • Clarence 'Kelly' Johnson designed the U-2 and is seen here (left) with Gary Powers, who would be shot down and put on trial in 1960, enhancing the urgency for a spy satellite programme. (Lockheed)

Precise identification allowed identification of 'known' targets within a matrix of other known types, and description provided detail for determining the dimensions of a target, proportionality of shape and the general layout or configuration of a building or a surface feature. Technical intelligence would provide a resolution sufficient to identify specific types of weapon and equipment or to identify specific military units, even the identifying features of unit markings on individual vehicles, ships, or aircraft.

There had been so little photographic coverage of the USSR from the U-2s that vast areas were largely unknown. For a long time, US intelligence agencies used captured German maps of Russia compiled from high-altitude photographs taken by the Luftwaffe during the Second World War. They revealed the incongruities in Soviet maps of Russian territories, deliberately falsified to confuse an invader. The job of mapping the USSR would have to be done from external sources and most of the work would be done by satellites.

Because of this escalating situation, in the 1950s information gathered by the CIA was increasingly seen as key to US State Department policies and satellite intelligence was regarded as a vital asset in determining the level of stability in Russia. That in turn impacted on negotiations on everything from food to arms talks.

Because of opposition in high places at the Pentagon, not until March 15, 1955, did the DoD issue its 'General Operational Requirement for a Reconnaissance Satellite Weapon System', GOR-80. As then projected, the satellite would have a radio-TV broadcast system as well as a signals intelligence (SIGINT) instrument for intercepting Soviet radio communications, with operational readiness of an initial precursor system set for mid-1960 and a fully operational capability by 1965.

Proposals sought

In the spring of 1955, design study proposals were sought from selected contractors. In wishing to maintain maximum security, only four companies were approached. Lockheed Aircraft Company, RCA and Glenn L Martin Company agreed to submit proposals, but Bell Telephone Laboratories declined. The three contractors began their work in June 1955 and submitted results in March 1956, each presenting a distinctly unique approach to the requirement, with Lockheed being favoured as the most attractive concept. Air Research & Development Command (ARDC) issued a system development directive, War (later Weapon) System-117L (WS-117L), on August 17, 1956, with Lockheed being awarded a contract on October 29.

By the summer of 1957, it had been decided that the original camera concept had to change to one in which a film scanning system would be employed. Film would be exposed and processed automatically on the satellite and a light beam would transform the pictures into electronic signals which would be transmitted when the satellite was over a ground receiver. It was felt to be the best compromise between real-time direct transmission and film return capsules plucked from the ocean after re-entry. In all of this, the Air Force was defining a satellite concept for which there was no precedent and no certainties.

In the rush to provide the military with a comprehensive suite of capabilities, WS-117L (at this date named Sentry by the contractors) embraced not only visual and electronic reconnaissance – and by implication, surveillance – but also infrared reconnaissance of weather systems and communications capabilities for maintaining global support of US forces. Briefing charts of the time include detection

BELOW • Determined to provide a more secure way of accessing military sites deep inside the Soviet Union, the CIA funded the Lockheed U-2 spy plane, the first of which was delivered in 1955. (USAF)

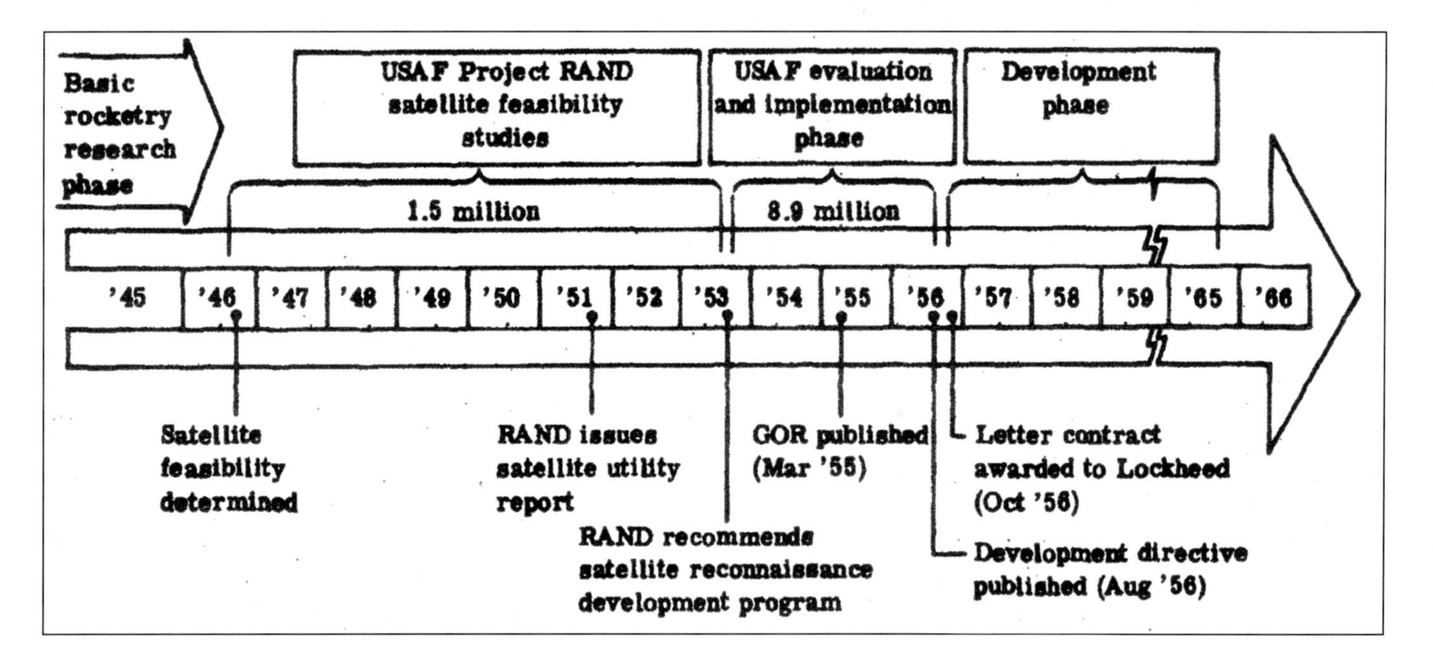

of nuclear tests, the use of Sentry for strategic warning of attack and the launch of enemy ICBMs. But much of this was wishful thinking on the part of Lockheed, keen to push their Agena rocket stage as the basis for an entire family of military satellites with wide-ranging applications.

Even at this early stage it was accepted that precursor development satellites would have an initial 'Pioneer' capability, leaving the more developed 'Advanced' system for later. The Pioneer system was postulated to provide a camera system with a 152mm (6in) focal length, a lens speed of f2.8 within a camera system weighing 136kg (300lb). Lockheed based its analysis on obtaining strips of exposed film covering a section of the Earth's surface 160.9km (100 miles) in width and 3,218km (2,000 miles) in length. The mission was expected to last nine days.

The more advanced system would utilise a camera with a focal length of 914mm (36in) at f2.8 covering an area 27.3km (17 miles) in width and 579km (360 miles) in length intermittently. This camera would weigh 181kg (400lb) and operate for 30 days. The wide-area Pioneer system would provide a ground resolution of about 300m (100ft) while the advanced close-look system would see objects down to 6.1m (20ft). Each camera system would operate on 300W of electrical power with a predicted 660W/hr per day. Film used was to be Microfile Eastman F570-6.

Advanced polar-orbiting TV satellites would operate from an orbit of 1,609km (1,000 miles) with direct readout to ground stations as they passed across the United States. The earliest recommendation for use of the Thor rocket was in a report released on November 12, 1957. It envisaged abandoning the WS-117L for this interim system, applied as a more permanent programme which it was believed could become the core of several sequential developments. The proposal for the Thor-launched concept was presented by Lockheed on January 6, 1958, a highly detailed description of an efficient spy satellite system.

Already involved in a range of proto-operational space ideas, General Electric suggested a camera subsystem, a recoverable film-carrying pod, a camera with an f3.5, 45.7cm (18in) lens and a resolution of 23m (75ft). GE's initial idea was to adopt a standard GE re-entry body of the type developed for ballistic missiles, dispatched to an ocean splashdown, whereupon the cone would crack open and allow the film-carrying sphere inside to float and be recovered by ships.

The break with WS-117L came on February 28, 1958, when the director of the Advanced Research Projects Agency (ARPA) sent a memorandum to the Secretary of the Air Force disconnecting this from the Lockheed satellite and adopting Thor boosters for a programme which, eventually, would be known as Discoverer, a concealed veil for photographic reconnaissance and biological experiments. It had already been decided the previous month to schedule nine Thor flights for the programme and later an additional five flights were authorised for biomedical science.

On March 24, 1959, ARPA approved the Discoverer programme. On December 3, 1959, ARPA transferred the Discoverer programme to the Air Force Air Research and Development Command. It closely examined what was internally named the Sentry programme and in June 1959 moved the mapping and charting capabilities to Samos, before moving the reconnaissance satellite to the Air Force on 17 November 1959. The name Samos was chosen by the ARPA director, Admiral John Clark, and refers to the Greek island, not 'Space and Missile Observation System', which countless writers and historians have assumed ever since.

This was also the date the Air Force took over the MIDAS (Missile Defense Alarm Satellite) programme from ARPA. In effect, WS-117L became three sub-programmes: Corona, MIDAS, and Sentry. By 1959, Discoverer was the

ABOVE • A page from the declassified history of the Corona spy satellite programme provides a timeline for its development with the General Operational Requirement (GOR) set in the time the first U-2s were delivered. (CIA)

BELOW • The initial Corona concept was built on the Agena rocket stage, which would be launched into space on a Thor rocket. The aft end of the stage (left) contains the single rocket motor, the cameras and recovery pod, or 'bucket' on the other end. (USAF)

public name for the Top-Secret Corona spy satellites. Samos represented three separate programmes: Project 101A (E2) for constant readout television spy cameras, Project 101B (E5) for film recovery of high-resolution imagery and Program 201 (E6) for film recovery and high resolution. MIDAS was for an infra-red sensor capable of monitoring missile launches by observing their hot exhaust plumes from space.

In some respects, Corona was put together to fast-track a spy satellite capability before the more advanced system could be deployed. Then the follow-on system was cancelled, and Corona endured for more than 14 years, transforming the way Western intelligence saw Russia's military potential and re-writing the text books on reconnaissance and surveillance.

Diversity and decisions

Between April and July 1958, essential elements of the programme were put in place, including the decision on launch schedules in April and the completion of a fabrication and assembly plan involving the Air Force, the CIA, Douglas Aircraft, Lockheed, Itek, General Electric and the Hiller Aircraft Company. Hiller had a small facility near Lockheed, and it was from them that Lockheed leased a building to assemble the Corona payloads to go on the Agena stages. Some Hiller people were hired by Lockheed, and it was there that the company founded its 'Skunk Works', but most were transferred to the new unlisted facility paid for through Hiller, who were told that the work was proprietary and could not be discussed.

To maintain contact between the organising and participating partners in Corona, the CIA set up a special cryptographic teletypewriter which linked together the CIA headquarters, the Ballistic Missile Division, and the Lockheed Skunk Works. Bogus mail drops under fictional names helped scatter mischievous trackers of communication links attempting to trace the connections, much of which was in hard copy. In the non-digital world of the late 1950s, these things were much easier to achieve than they are today. However, some people knew too much simply because of their background.

One official Air Force officer who had previously been made aware of WS-117L, when hearing of Discoverer, tried energetically to persuade managers that they should be looking to use the satellite for spy flights rather than the scientific investigations publicly declared, without realising that beneath Discoverer lay Corona, unseen and on a similar function. It served its purpose to have him associated with the Discoverer programme because those who knew his background could believe that if he didn't know about a spy satellite programme, it couldn't possibly exist!

While upper stages had been developed for other rockets, including the Vanguard IGY satellite launcher, the development of Agena was unique. The name was chosen by an ARPA committee sometime around mid-1958 and that fitted the Lockheed tradition of naming its products after celestial phenomena. Also named Beta Centauri, Agena is the tenth brightest star in the sky – and that seemed appropriate enough.

The first Agena A stages employed the Bell XLR81-BA-3 (Bell Model 8001) rocket motor before moving to the YLR81-BA-5 version and had a gimballed nozzle for pitch and yaw control during ascent. The BA-3 motor delivered a thrust of 68.9kN (15,500lb) for up to 120 seconds. Attitude control gas jets were used for yaw stabilisation. The BA-5 had a specific impulse of 277secs (defined as the amount of thrust delivered for one second by one kilogramme of propellant) and a thrust of 69.40kN (15,600lb). The Agena A tanks carried a total propellant load of 2,960kg (6,525lb).

Agena A had a total length of 5.94m (19.5ft) with a diameter of 1.52m (5ft) and a total all-up weight of 3,850kg (8,500lb), although some of the later ones pushed up to 3,930kg (8,662lb). In this condition, the stage carried a payload of 225kg (497lb). The Thor rocket used for the Agena A (DM-18) had a lift-off thrust of 676.1kN (152,000lb) with a stage burn time of 163secs.

With longer propellant tanks, Agena B had a length of 7.56m (24.8ft) but the same diameter as Agena A and a fuelled weight of 7,160kg (15,800lb). Early examples of the Agena B employed the XLR81-BA-7 rocket motor (Bell 8081) with a burn time of 240 seconds, double that of its predecessor. Later models employed the XLR-81-BA-9 (Bell 8096), which had a slightly higher thrust, at 71.1kN

ABOVE • *Corona satellite retrieval involved a recovery aircraft such as this Lockheed JC-130B, demonstrating how a descending parachute could be snared and the capsule reeled in. (USAF)*

(16,000lb). Agena B had a dual re-start capability and was carried aloft by a Thor DM-21, lighter and with the more powerful Block 2 MB-3 rocket motor delivering a thrust of 751.7kN (169,000lb).

Mating possibilities

The evolution passed to Agena D, essentially the same as the B but capable of mating to either a Thor, Atlas, or Titan first stage without modification. It also had provision for a conical payload section in the nose and this greatly assisted payload planners integrate with the highly versatile upper stage. NASA would use Agena D for many of its missions and adapted it as the target vehicle for the Gemini flights of 1966, its restartable motor pushing Gemini to record altitudes for Earth orbiting manned spacecraft which stand to this day.

Several code names were applied to Corona. Using familiar methods of coding, a two-word name implies a secret programme with restricted access but an accessible identity while a single word implies a Top-Secret piece of hardware, in this case a specific satellite, or series effectively using the same lineage. Thus, it was that the programme name can be found identifying Corona as being within Operation Bootstrap or Project Forecast, while the name of the satellite itself remained constant.

In reality, Corona embraced a separate set of sequentially more capable systems in three versions carrying two camera packages as selectable payload options. Initial flights would carry a single camera/single recoverable capsule, followed by a dual stereo camera/single capsule or dual stereo camera/dual capsule design. They would all be embraced by the Corona programme but at various times the names changed for each of the three selectable types. Other names attached would be Mural (triple stereo camera version), Argon or Lanyard.

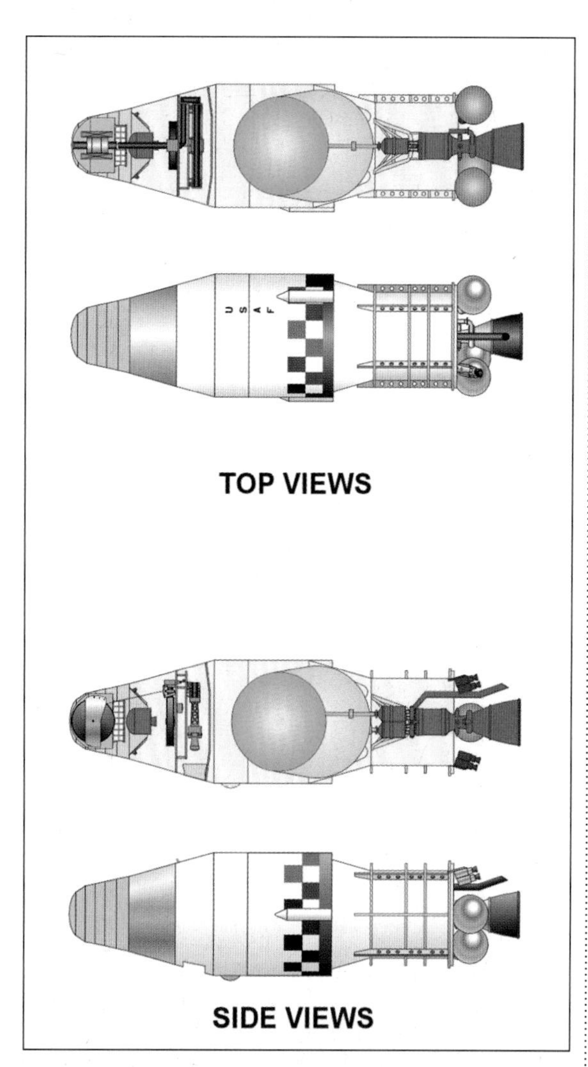

LEFT • The first-generation Corona, the KH-1, carried a single recovery pod delivering the film back to Earth. (Giuseppe De Chiara)

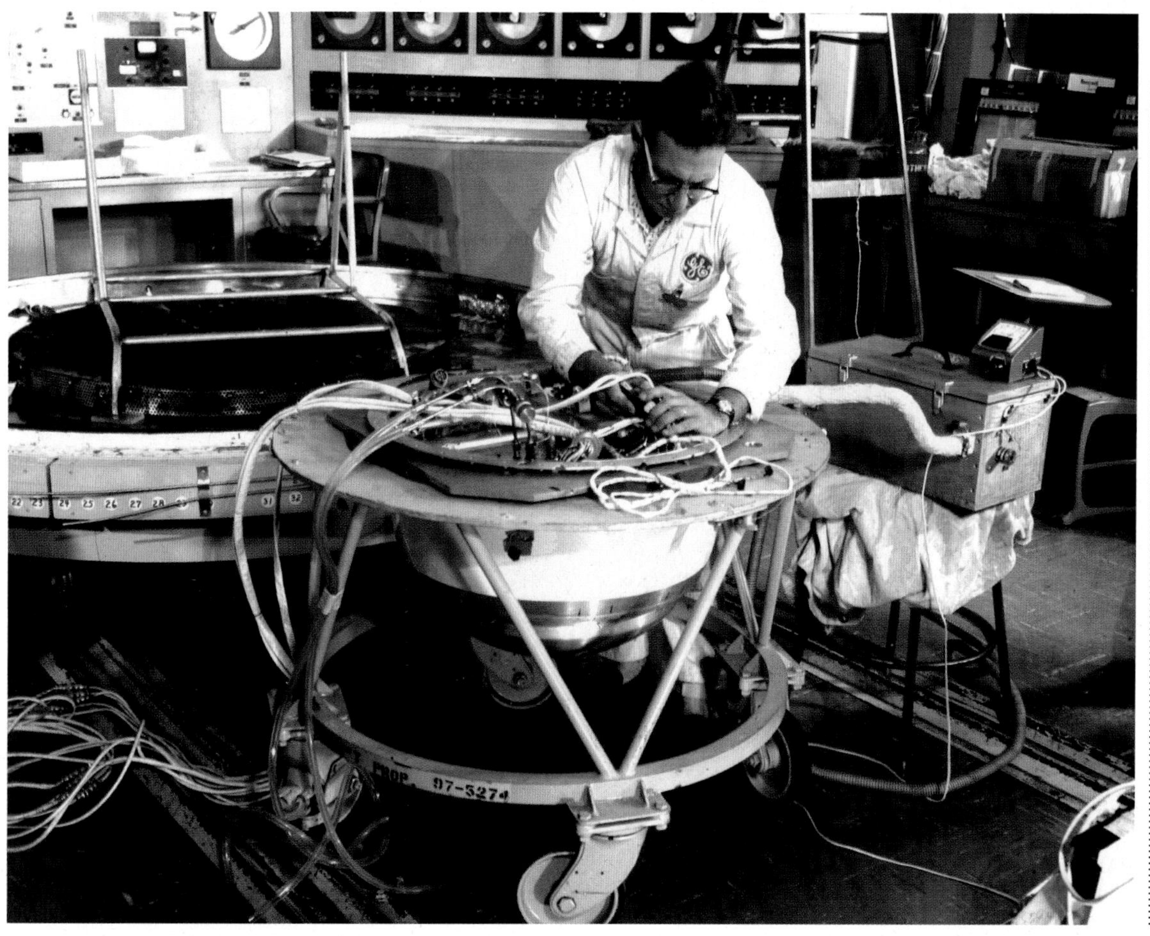

LEFT • A Corona capsule of the Discoverer programme being equipped with biological experiments, validating the public explanation that this was a purely scientific research endeavour. (USAF)

For security classification purposes, another parallel coding system was adopted using the two words Key Hole (KH), from the Keyhole Byeman system: KH-1 for C; KH-2 for C', KH-3 for C''' and KH-4 for Mural systems; KH-4A for Corona J-1; and KH-4B for Corona J-3. Taking its name from a person who works underground – a bye man – when it was set up in 1961 to run the US spy satellite programmes, the National Reconnaissance Office adopted the Byeman Control System (BCS) configured a year earlier by the CIA. More commonly known simply as Byeman, it was phased out in 2005 and replaced with the Talent-Keyhole Control System (TKCS).

That name was derived by combining classifications for highly sensitive photographs obtained from overflights of Soviet territory by manned aircraft, set up in 1956, (Talent) and for information obtained from reconnaissance satellites (Keyhole), from 1960. The Byeman system protected the development of the satellites while the Talent and Keyhole systems controlled access to the data they produced, the latter folding into the former after 2005. Further, but classified, extrapolations have been introduced, making interpretation similar to 3D chess games, where three players each control one play in three planes simultaneously.

What had at the beginning been a rather informal effort closely guarded within the appropriate channels at the Air Force and the CIA became increasingly formal and institutionalised. Established on September 6, 1961, the National Reconnaissance Office (NRO) remained secret for three decades and became the home of the reconnaissance satellite. The first director of the NRO, Dr J V Charyk, chose to leave Corona within the CIA, since it was deemed to have a limited life remaining and would soon be superseded by SAMOS E-5. But the Lanyard system was moved across to the NRO.

Bucket test

An operational Corona flight would incorporate a General Electric Satellite Recovery Vehicle (SRV), commonly known as the 'bucket' because it would be released from the forward section of the Agena and return to Earth with the film shot during orbital operations. Launched in a southerly direction from Vandenberg Air Force Base, California, the spacecraft would enter a polar orbit, flying first over the South Pole. At the end of its nominally one-day mission, the ejection sequence for the bucket would be initiated by radio command from the station in Alaska and recovery would be conducted by an airborne snatch.

One drawback with the use of Vandenberg was that the Southern Pacific railroad ran close by, and the Air Force was nervous about having car loads of passengers gawping at the launch site with cameras clicking while launching what only a few knew were highly classified Corona payloads masquerading as scientific research flights. By scrutinising the timetables, launches could be synchronised to windows within gaps in the Southern Pacific schedule. Not that there was anything to see but so tight was the security that it just didn't seem right to have trainloads of inquisitive passengers rolling by!

Tied to this launch window was a requirement to launch in daylight and to synchronise the path so that on a typical 24hr mission of 17 orbits, seven would cross Soviet and Chinese territory, also allowing for daylight recovery. By combining these requirements, an afternoon launch was the optimum period in the day – with the precise time set by the Southern Pacific timetable!

After the photo sequence was complete and the command was sent to de-orbit the SRV, the Agena stage would pitch down 60° so that the bucket would be in the correct orientation with respect to the flight path. After the SRV separated from the front of the assembly, it would be spin-stabilised at 70.5rpm by small spin-rockets to maintain it at that angle, the nose section pointing down. The retrorocket would fire to decelerate the capsule by 387m/sec (1,270ft/sec) some 3min 27secs later and de-spin rockets would fire to slow the rate of rotation to 10.5rpm. The retrorocket thrust cone would separate from the blunt aft end and the SRV would begin re-entry.

During descent, the nose cone would decelerate due to friction with the atmosphere, kinetic energy released as heat being removed through ablation. About 9min 37secs after thrust cone ejection, the heat shield would fall away to reveal the spherical container – the bucket – with the exposed film. This would release a drogue parachute followed by the main parachute in a reefed condition, fully deployed 5secs later to lower the capsule to the recovery area.

The recovery technique called for the aircraft to fly over the descending capsule and snag the parachute or its shrouds in a trapeze-like hook suspended below the aircraft and winch the capsule aboard. During the test phase before flights began, only 49 recoveries were achieved in 75 simulated drops using personal parachutes. Using an early type of recovery parachute, only four were recovered

BELOW • Corona satellites were launched into polar orbits from Vandenberg Air Force Base, California, with this Thor rocket being prepared for such a flight in early 1959. (USAF)

of 15 attempts but after the design was changed, there were five successes in 11 drops.

The real problem was in the sink rate of 10m/sec (33ft/sec) with a parachute weighed down by the mass of the capsule. The ideal descent rate was 6m/sec (20ft/sec), during which a recovery aircraft would have time for four passes to get it right. Early flights would use the Fairchild C-119 but later the Lockheed C-130 Hercules was inducted for bucket recovery. By the time space flights started and bucket capture was required for real, the parachute design had improved markedly, giving the pilots the option of two or three passes.

Flight time

The Discoverer programme reached flight status in little more than a year after departing from the mainstream development of WS-117L. The first two scheduled Discoverer flights did not have buckets because they carried no cameras. The first flight was set up to go from LC-75 at Vandenberg, two years after the first flight of a Thor missile with 30 flights, of which 17 had been failures. Nevertheless, the first two-stage configuration, the Thor-Able, had demonstrated that the rocket could carry an upper stage into orbit.

The first attempt on January 21, 1959, was designated Discoverer 0 and was aborted due to a spurious electrical signal causing a near catastrophic explosion and the first flight occurred (as Discoverer 1) on February 28. Due to a low injection angle, there was some uncertainty as to whether the Agena A had made it into orbit as no radar tracking was obtained and no telemetry signals were received. For several hours, tracking stations in Alaska and Hawaii tried to raise it.

This was an exacting and drawn-out process. As the Earth rotated on its axis at 1,670kph (1,037mph), with each 90min orbit the ground track of the spacecraft would appear to move approximately 2,500km (1,550 miles) further west each time it crossed the same spot on the equator. Despite successive bids to acquire a signal, and some reports that a faint response had been detected, an investigation determined that Discoverer 1 had re-entered over Antarctica before achieving its planned polar orbit.

Launched between April 13, 1959 and June 29, 1960, the next 12 attempts were unsuccessful, the first SRV recovery occurring with the Discoverer 13 mission launched on August 10, 1960, the first man-made object returned from orbit. This was followed by the first fully successful flight of Discoverer 14 on August 18, recovery of the film capsule a day later on the third pass by a C-119 aircraft. Discoverer 15 was launched on September 13, but the bucket was not recovered. It was the last of the C-series (KH-1) cameras.

Launched on October 26, Discoverer 16 carried the first of the C' (KH-2) cameras but success was only achieved by Discoverer 18, launched on December 7, 1960. KH-2 had almost twice the amount of film as the KH-1, covering more than 9,842,000km^2 (3,800,000 $miles^2$) of Soviet territory. With 65 lines/mm compared to 55 lines/mm for KH-1 cameras, resolution was 20 per cent better, the pictures showing some objects as small as 7.6m (25ft). The satellite remained in space for three days, versus the previous one day for KH-1 flights, and that allowed the gradual migration of the orbital ground track further west with each pass, successive slices of territory coming into view.

Overdone extravagance

Although KH-2 was not a successful programme, in seven attempts to August 4, 1961, only one was an unqualified success when Discoverer 18 disclosed the low level of Soviet ICBM, triggering realisation that the United States was far ahead of the USSR in strategic nuclear weapons. Bravado and extravagant claims bolted to a closed society had overdone the job of putting fear into the hearts of potential enemies when, in reality, Soviet propaganda campaigns had been baseless and untrue. Space spectaculars had created a false canopy of alarm and over-reaction.

Launched on February 17, 1961, Discoverer 20 carried the first KH-5 Argon mapping system using a frame type camera built by Fairchild. It had a line resolution of 3.5mm (0.14in) per scan on to 3404 Estar film, 12.7cm (5in) wide

ABOVE • *In April 1964, Venezuelan agricultural workers got hold of a Corona camera film bucket when it came down in a remote part of the country and brought visitors from miles around to see the strange object before US government officials were alerted and retrieved it. (NRO)*

ABOVE LEFT • *Discoverer 13 returned this capsule proudly displayed by (from left) Col 'Moose' Matheson of the 6954th Test Wing, Maj Gen Osmond Ritland of the Air Force Ballistic Missile Division, Lt Col Bernard Schriever of the Air Research Development Division and two civilians. (USAF)*

ABOVE • *Spy satellite operations fed directly into aerial reconnaissance operations, including those of the Lockheed SR-71 Blackbird. (USAF)*

shot through a 7.6cm (3in) focal length lens. KH-5 was also equipped with a 7.6cm stellar camera. Discoverer 30 failed and the first successful Argon was not before May 15, 1962. Across 12 launches, only six were successful, the last on August 21, 1964.

Covering an area on the ground 556km x 556km (345 miles x 345 miles) with a resolution of 140m (459ft), KH-5 satellites were suitable for wide-scale mapping of very large areas and as such provided the Army with valuable and relatively detailed photographs. The spacecraft weighed an average 1,300kg (2,866lb) including the Agena B stage, the mass varying according to the specific piggy-back payloads and operating life ran to a few days in orbit.

Attempts to get the KH-3 up and running came on August 30, 1961, with Discoverer 29, but an incorrect scan head failed to focus the camera. After five more failures, the first success began with the launch of Discoverer 35 on November 15, repeated with Discoverer 36 on December 12. But these were the only successful KH-3 missions, the last of nine launched on January 13, 1962, also failing.

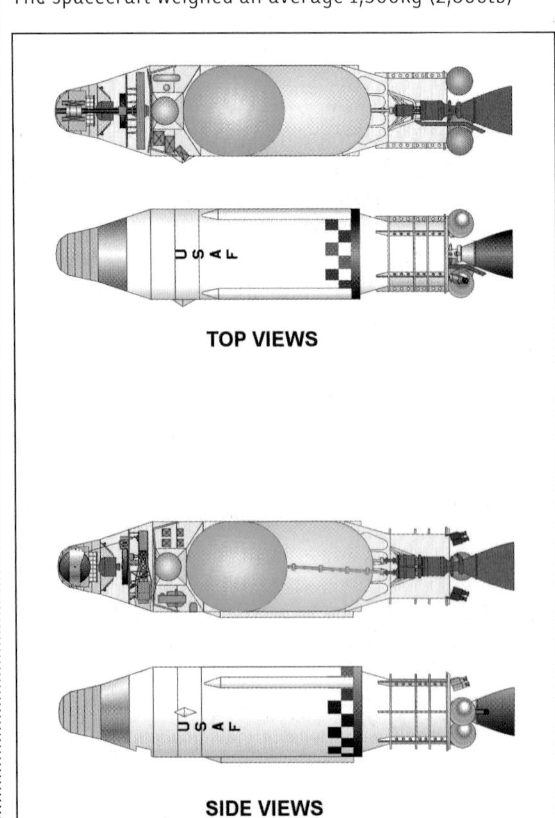

RIGHT • *The KH-2, C' variant of the Corona programme adopted the Agena B upper stage configuration. (Giuseppe De Chiara)*

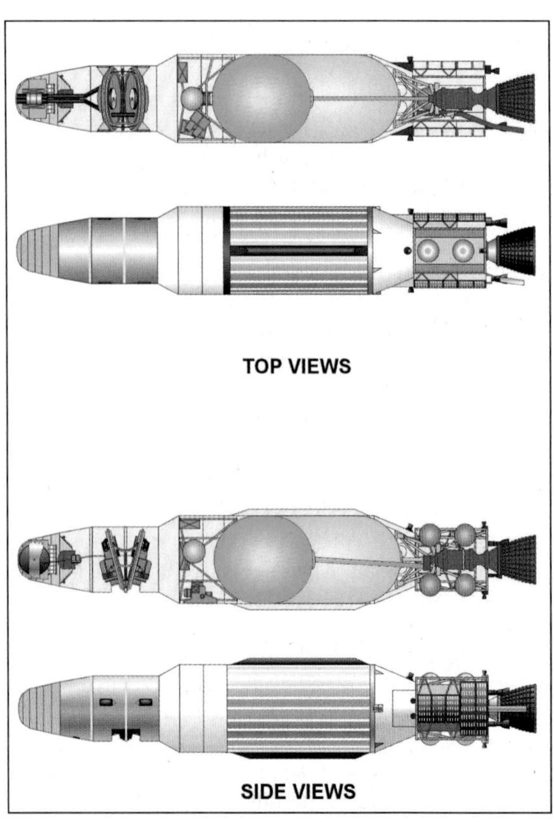

FAR RIGHT • *The KH-4 Corona variant was the first to use the Agena D, which had significant potential through its restartable rocket motor, allowing it to change orbit according to operational requirements. (Giuseppe De Chiara)*

The 61cm (24in) f3.5 lens for the KH-3 was of the Petzval type (versus the f.50 Tessar for the KH-1 and KH-2), with a 70-degree pan. It shot to a 3400 Estar film with 2.5mm/line (0.1in/line) scan producing a resolution of around 11m (35ft), compared with 12m (40ft) for the KH-2 and 15m (50ft) for the KH-1. At 2,377m (7,800ft) the film supply was the same as the KH-2. The technical capabilities of the C''' camera were well proven through the two highly-valued missions in which it was a great success and integral to the next development.

The first KH-4 was launched as Discoverer 38 on February 27, 1962. It was a development of the C''' but carried the title Mural because of the significant change to its operating protocols. Much like Corona had begun development as a quick-reaction interim project until a film read-out system could be launched under the original expectations of WS-117L, so too was Corona-M expected to be an interim programme to take existing equipment (the C''') and provide stereo camera capability.

In addition to the twin C''' cameras, an Itek index camera was carried on this first KH-4 flight. It had a downward-looking terrain lens of 38mm (1.5in) but the stellar index camera on later flights had both a 38mm lens and an 80mm (3.1in) focal length stellar lens; the former using 70mm (2.75in) film, the latter using 35mm (1.37in) film. The dynamic resolution was 80-110mm/line (3.1-4.3in/line).

Missions carrying only the index camera ran through to the 12th flight launched on 29 August 1962, but the second flight (Discoverer 39) was the last to bear that name. The performance record was increasing significantly. By the end of June 1962, a total of six KH-4 satellites had been orbited with the loss of only one capsule. What had begun as an interim programme had become highly successful and would extend for several more years, the last Mural system being launched on December 21, 1963, completing 26 successful launches and 20 recoveries.

Designated KH-4A, the next evolution utilised the more powerful Agena D launch system with two spools of film, each feeding a single SRV and with a quiescent mode, allowing up to 21 days between operations. This would allow the orbit to migrate, or to be adjusted with the Agena D restart capability, so as to visit the same area several times. Or it could go quiet in space, progressing its orbit until required. The satellite could appear to have either returned its recovery capsule, or to have malfunctioned, only turning itself on by command when required.

Known as Corona-J for Janus, the Roman god of transitions and two faces, the KH-4A had capacity for 9,750m (32,000ft) of film, first passing through the first SRV and out to the second capsule. When that was full, a cutter would sever the film thread, sealing the SRV for return to Earth, the second SRV located behind the first receiving exposed film and eventually returning to recovery. No one individual was allowed to see the entire mechanism. Each component and unit would be manufactured separately and only brought together when the assembly process had itself been hidden within the overall design, thus avoiding the possibility of any one individual being able to visually reverse engineer the design.

Success and failure

The first KH-4A was launched on August 24, 1963 and returned the first capsule but the second pod failed to separate due to electrical failure. The second flight suffered the same fate but the third, launched on February 15, 1964, was a resounding success. The fifth flight launched on April 27, 1964, had a curious fate. Seemingly struck by a series of technical problems, the Agena D degraded naturally back down through the atmosphere, providing a fiery spectacle over the mountains of southwest Venezuela, the remnants striking the ground outside the town of La Fria.

Retrieving one of the buckets, two farm labourers brought it to the farmer who disassembled the capsule for toys to occupy his children and to sell remnants to a growing trickle of local residents tramping the countryside to see what he had found. A commercial photographer saw it and called the US Embassy in Caracas, whereupon officials visited the farmer and bought it from him, explaining that it was from a NASA satellite and of no value but only scientific interest.

News broke and US newspapers reported the story with apparent disinterest and lack of attention to its true purpose, which they did not in fact know. The reason for the lack of interest? Earlier that same day the press was anxiously reporting the attack on two US warships by North Vietnamese patrol boats in the Gulf of Tonkin. The Vietnam War had begun.

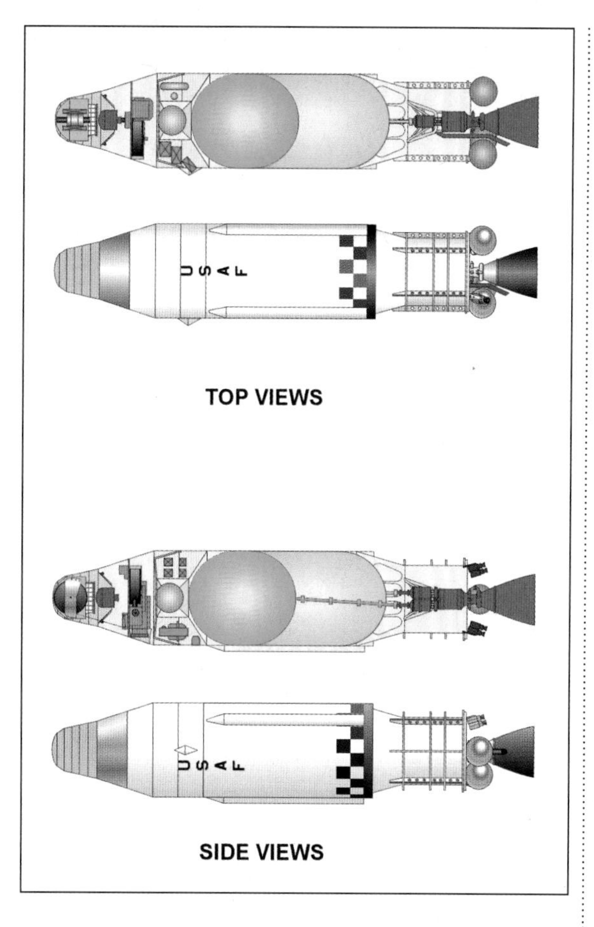

LEFT • The KH-5 ARGON followed in 1961, as shown here on an Agena B, a mapping camera system much sought by army cartographers. (Giuseppe De Chiara)

BELOW • This display model of a Corona shows the main camera and optics section to the left and the recoverable film pod ('bucket') with a tan-coloured strip of film connecting the two sections. (Author's collection)

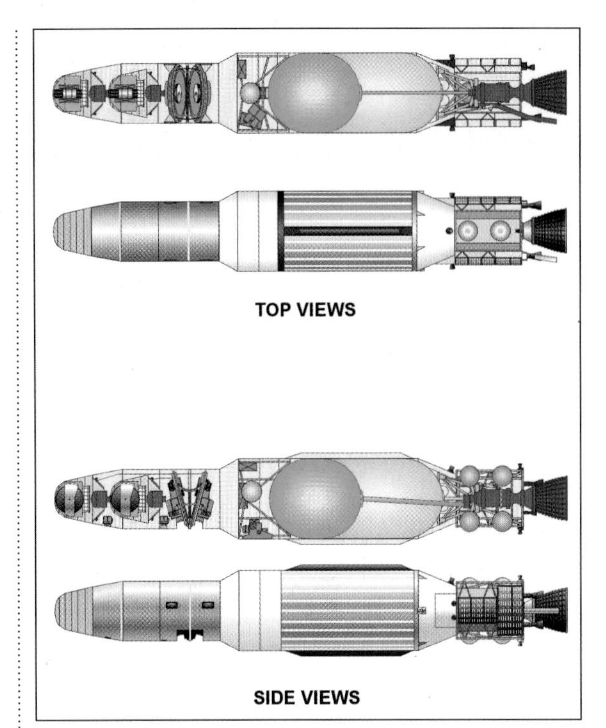

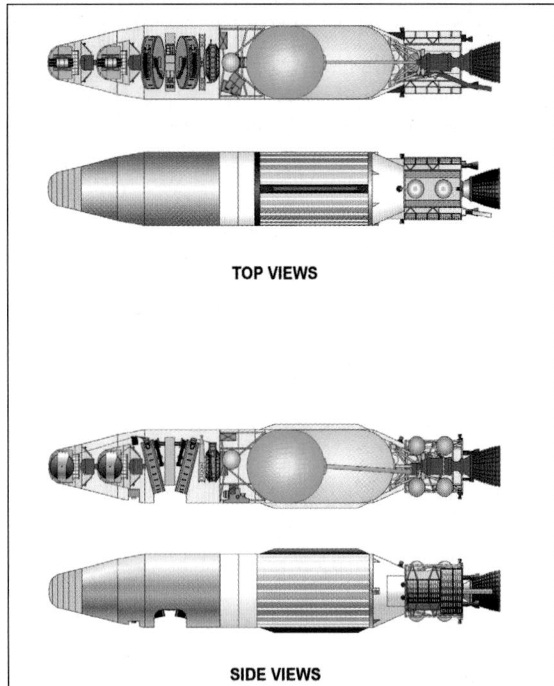

RIGHT • With its restart capability, the Agena D multiplied the mission options for Corona satellites and significantly extended the duration of each flight. (Giuseppe De Chiara)

FAR RIGHT • Incorporating a rocking camera system, the KH-4B was a further development which enhanced operating flexibility and produced improved images. (Giuseppe De Chiara)

Better eyes

Expanded requirements and demands for better and bigger satellites resulted in the KH-4B system. Requested by the Air Force Space Systems Division in January 1966, the basic Thor stage, which had remained largely unchanged throughout, was extended in length, allowing it to carry additional propellant, with the forward fuel tank having a constant diameter rather than tapering as before. Added to this were improved and more powerful strap-on solid propellant booster rockets.

Known as Thorad-Agena D, the first of these carried a KH-4A into orbit on August 9, 1966. The new launcher could lift 1,315kg (2,900lb) into polar orbit, more than 300kg (661lb) greater than the TAT-Agena D, allowing a weight increase for the final Corona development, the KH-4B, or Janus-3. It was improvements to the operation of the cameras that gave the KH-4B its distinctive edge and produced the truly definitive Corona.

Previous camera systems had operated where the lens assembly rotated as the shutter unit and the slit oscillated. This rocking motion caused vibration when the combination came to a halt, reversed direction, and started again. This had an effect on the stability of the Agena and on the quality of the photographs. The revised configuration was

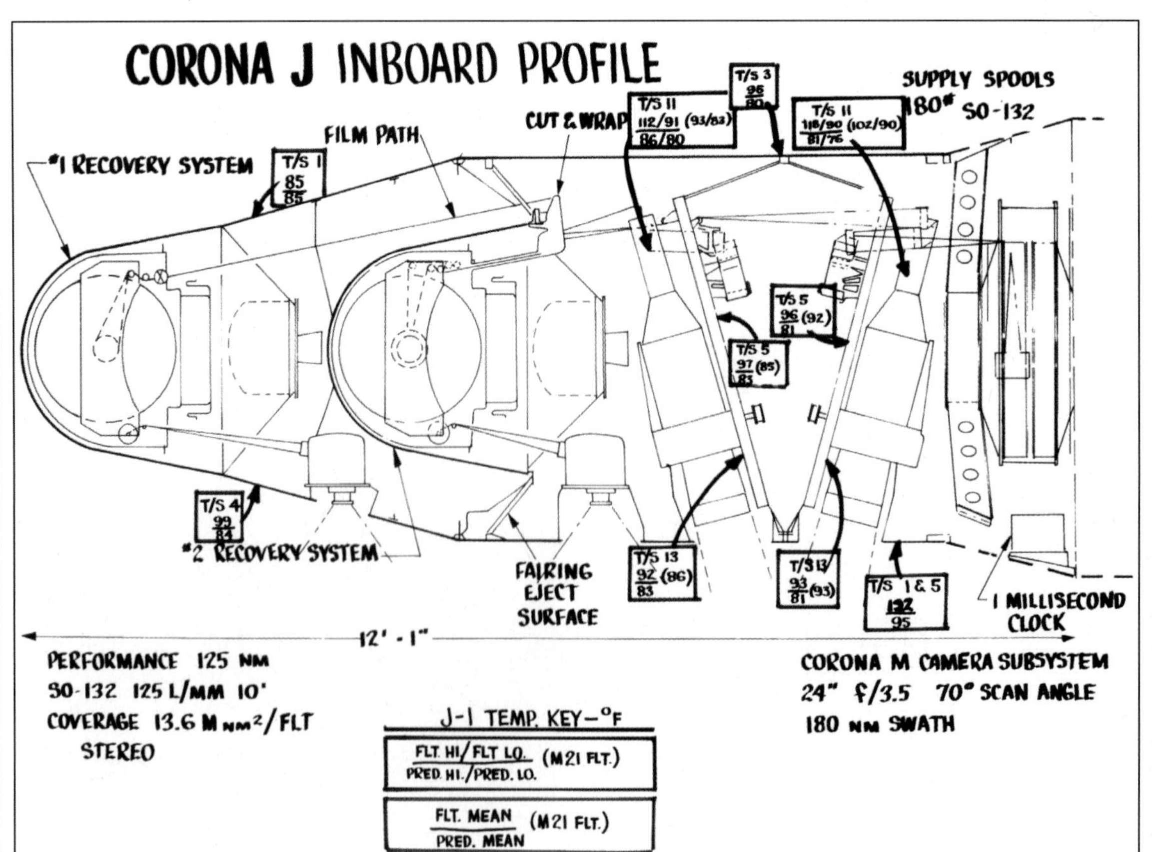

RIGHT • The addition of a second recovery 'bucket', the KH-4A (Corona J) and the extended life of the Agena D to which it was attached, effectively providing two operational cycles for one launch. (NRO)

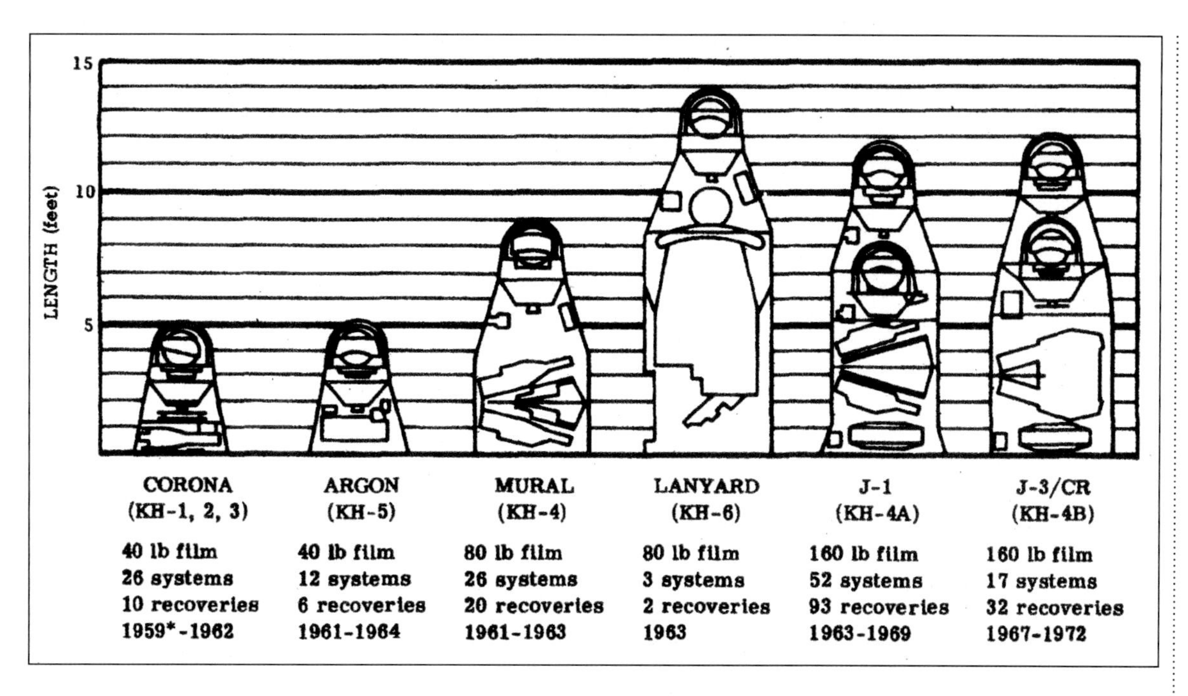

LEFT • Between 1959 and 1967, significant improvements were made to the Keyhole camera equipment and design of the optical and satellite operating systems. (National Reconnaissance Office)

called a 'constant rotator'. A rotating drum incorporated the lens assembly, shutter and slit with the film only exposed during a 70 degree of arc rotation.

To start filming, the two separate drum assemblies would rotate in opposing directions and to compensate for the orbital motion of the satellite, the two separate camera assemblies would rock back and forth, also in opposite directions. This would neutralise gyroscopic motion from the mass of the drums, which were each over 1.5m (5ft) in diameter and 18cm (7in) deep. Resolution was increased to 160lines/mm. These refinements improved the ground resolution to 1.8m (6ft), compared to 2.7m (9ft) for the KH-4A.

The first KH-4B was launched on September 15, 1967, the 120th flight of a Corona satellite lasting 19 days and with recovery of both SRVs. As the programme progressed, various film emulsions were tried out, including some infra-red and colour film as well as special thin-film products for later programmes. Albeit with lower resolution, colour allowed analysts to distinguish between dry and wet land and between natural vegetation and managed land in a way impossible with monochrome film.

The last KH-4A was launched on September 22, 1969, the 135th Corona. The KH-4A had taken reconnaissance photographs covering 707,189km^2 (273,047miles^2), compared to 386,577km^2 (149,258miles^2) by KH-1, KH-1, KH-3, and KH-4. In addition, it had covered 97,149km^2 (37,509miles^2) of territory with mapping function. The last KH-4B was sent up on May 25, 1972, coverage totalling of 656,959km^2 (253,653miles^2) for reconnaissance and 98,509km^2 (38,034miles^2) for mapping purposes.

BELOW • Payload capabilities of the Corona series with optical types and basic technical specifications. (National Reconnaissance Office)

Designator	C (KH-1)	C' (KH-2)	C''' (KH-3)	ARGON (A) (KH-5)	MURAL (M) (KH-4)	LANYARD (L) (KH-6)	JANUS (J-1) (KH-4A)	J-3/CR (KH-4B)
Camera manufacturer	Fairchild	Fairchild	Itek	Fairchild	Itek	Itek	Itek	Itek
Lens manufacturer	Itek	Itek	Itek	Fairchild	Itek	Itek	Itek	Itek
Design type	Tessar, 24 inch, f/5.0	Tessar, 24 inch, f/5.0	Petzval, 24 inch, f/3.5	Terrain, 3 inch; Stellar, 3 inch	Petzval, 24 inch, f/3.5	Hyac, 66 inch, f/5	Petzval, 24 inch, f/3.5	Petzval, 24 inch, f/3.5
Camera type	70° pan, vertical, reciprocating	70° pan, vertical, reciprocating	70° pan, vertical, reciprocating	Frame	70° pan 30° stereo, reciprocating (2)	90° pan, 30° stereo, (roll joint)	70° pan, 30° stereo, reciprocating (2)	70° pan, 30° stereo, rotating (2)
Exposure control	Fixed	Fixed	Fixed	Fixed	Fixed	Fixed	Fixed	Slits (4) selectable
Filter control	Fixed	Fixed	Fixed	Fixed	Fixed	Fixed	Fixed	Filters (2) selectable
Primary film (film/base)	1213/ acetate 5.25 mil*	1221/ acetate 2.75 mil	4404/ estar 2.5 mil	3400/ estar 2.5 mil	4404/ estar 2.5 mil	3400/ estar 2.5 mil	3404/ estar 2.5 mil	3404, 3414/ estar 2.5 mil
Recovery vehicles	1	1	1	1	1	1	2	2
Subsystem (stellar/index)	None	None	None	N/A	1 S/I,† 80-mm stellar 38-mm terrain	1 S/I, 80-mm stellar 38-mm terrain	2 S/I's, 80-mm stellar 38-mm terrain	DISIC (Fairchild) 3-inch stellar (2) 3-inch terrain

* Support thickness
† Index only missions 9031-9044

BIGGER EYES

The need for high resolution imagery soon becomes apparent and the race for detail is on.

RIGHT • As the technical capabilities increased and the Agena D and its payload became heavier, operators switched to the Atlas to get the Keyhole satellites into orbit, as with this KH-7 during launch from Vandenberg Air Force Base, California. (USAF)

BOTTOM RIGHT • A KH-7 single-bucket spy satellite splendidly preserved at the USAF Museum reveals the details of its attachment and of its optical section adjacent to the roll joint. The Agena D is not displayed. (USAF Museum)

While Corona had provided general photographic evidence of area surveys and some detailed locations in unfriendly countries, there was an increasing need for high resolution imagery. For that, it was necessary to build a bigger satellite which required a larger launch vehicle and instead of the Thor-Agena used for Corona, the Atlas-Agena would be available for the bigger system. In several respects, while Corona was successful in parts, it was not the overriding success envisaged. But the evolution for the next generation originated even before Corona flights achieved success.

The need for a very high-resolution photographic system emerged in March 1960 when Eastman Kodak sent the CIA a proposal for permission to develop a camera with a 195cm (77in) focal length and three months later a 91cm (36in) system for stereo imaging on film. Known as Blanket, in July 1960 Eastman proposed a further development integrating the 195cm camera with stereo features and importing a film recovery technique from Corona. Eventually, this would be known as KH-7 Gambit but delays in getting it ready would tip planners to the interim KH-6 Lanyard.

KH-6 was a disappointment. It was quickly put together when CIA analysts believed they had spotted Soviet anti-ballistic missile facilities outside Leningrad. Using a previously rejected E-5 camera developed for Samos, with a focal length of 1.67m (5.5ft), it was capable of discriminating objects 1.8m (5.9ft) in size. Launched by Thor-Agena D, it would cover a swath of ground 14km (8.7 miles) by 74km (46 miles) across 910 frames on 686m (2,250ft) of film. The 1,500kg (3,300lb) payload supported a single re-entry bucket. Three were launched between March and July 1963 and only the last was moderately successful, albeit with a resolution no better than the KH-4B.

The key to getting the required imagery was to have a much more stable platform and here the design of the

KH-7 departed significantly from Corona. The total Gambit system had a length of 4.5m (15ft), a diameter of 1.5m (5ft) and a payload weight of around 523kg (1,154lb). It consisted of the Camera Optics Module (COM), the Orbital Control Vehicle (OCV) built by General Electric and the Satellite Recovery Vehicle (SRV), also built by GE. The weight of the total vehicle above the Atlas-Agena D was in the order of 2,000kg (4,410lb), the maximum potential lift capacity.

The COM was effectively three cameras in one assembly, a strip camera, a stellar camera, and an index camera. The strip camera accepted an image reflected from a 1.21m (4ft) diameter concave mirror which reflected light through an opening in a flat mirror and through a corrector.

Stereo operation would be achieved with a 30-degree included angle between pointing forward and then aft for an overlap. Continuous strip photography would adopt the same angular sweep between lateral pairs. Precision was vital as the optics would be looking at a slant angle with a total ground swath only 6.3° wide, where slight shudder or jitter would cause image distortion.

Continuous exposure

The strip camera continuously exposed a narrow strip on the film as the camera passed over the area being photographed. On the KH-7, it photographed a small target area on the ground through this narrow slit which was located close to the focal plane of the camera. This could produce stereo pairs, lateral pairs and strip photographs up to a maximum of 600 stereo pairs or an equivalent amount of continuous strip photography. It provided image motion compensation by moving the film across the exposure slit at the same velocity that the projected image moved over the Earth.

Gambit would carry 915m (3,000ft) of film 24cm (9.5in) wide. The camera would image a strip 10.6nanometres wide and the total payload system had a weight of 523kg (1,154lb). The format size was 22cm (8.7in) wide and of variable length up to 983cm (387in) but usually within the range 33-117cm (13-46in). Film speeds varied from 5.13cm (2.022in) per second to 9.61cm (3.784in) per second in 64 discrete segments. Stereo pairs would produce 100 per cent overlap while pairs of lateral photographs would produce parallel strips with minimal overlap.

An innovation borrowed from the Lanyard programme married the optical elements directly to the Agena stage, a coupling known as Hitchup. This introduced the 'roll-joint', a separate interface placed between the Agena D and the optical section which could take over attitude control and fine-pointing, achieving higher angular precision than was believed possible with the standard Agena D. It allowed the Agena to remain stable, leaving the forward section to roll from side to side for direct imaging, obviating the need for the entire system to change attitude.

A total of 38 KH-7/Gambit satellites were launched between December 12, 1963 and June 4, 1967, not all of which were successful as the new technology proved troublesome and prone to failure. While the Soviets were building their forces in a way that demanded high resolution monitoring, only the less capable KH-4A series was holding the line, albeit with good results. It was quickly discovered that quality control and a flawed contract which profited companies by delivering on time and under budget instead of performance and success was the core problem. The incentive was switched to reliability and performance improved.

Flights were sent up from a new launch complex at Point Arguello, part of the Vandenberg Air Force base, from 1964 and 34 successfully returned buckets, of which three had unusable film. Problems with early launches were not considered terminal but the need for an even better system resulted in Gambit-3, the KH-8 which had been under development since the early days of KH-7 operations. The demand from the intelligence community was outstripping the capabilities of existing satellites and the technology was leaping far ahead of the time required to organise, implement, and introduce it operationally.

An embarrassment of riches

The intelligence community wanted to achieve the highest resolution possible for area surveillance, defined as the need to continue investigating in depth targets that had been discovered from previous reconnaissance missions. By the

ABOVE • The optical cut-out on this declassified image of a GAMBIT system is clearly visible. (National Reconnaissance Office)

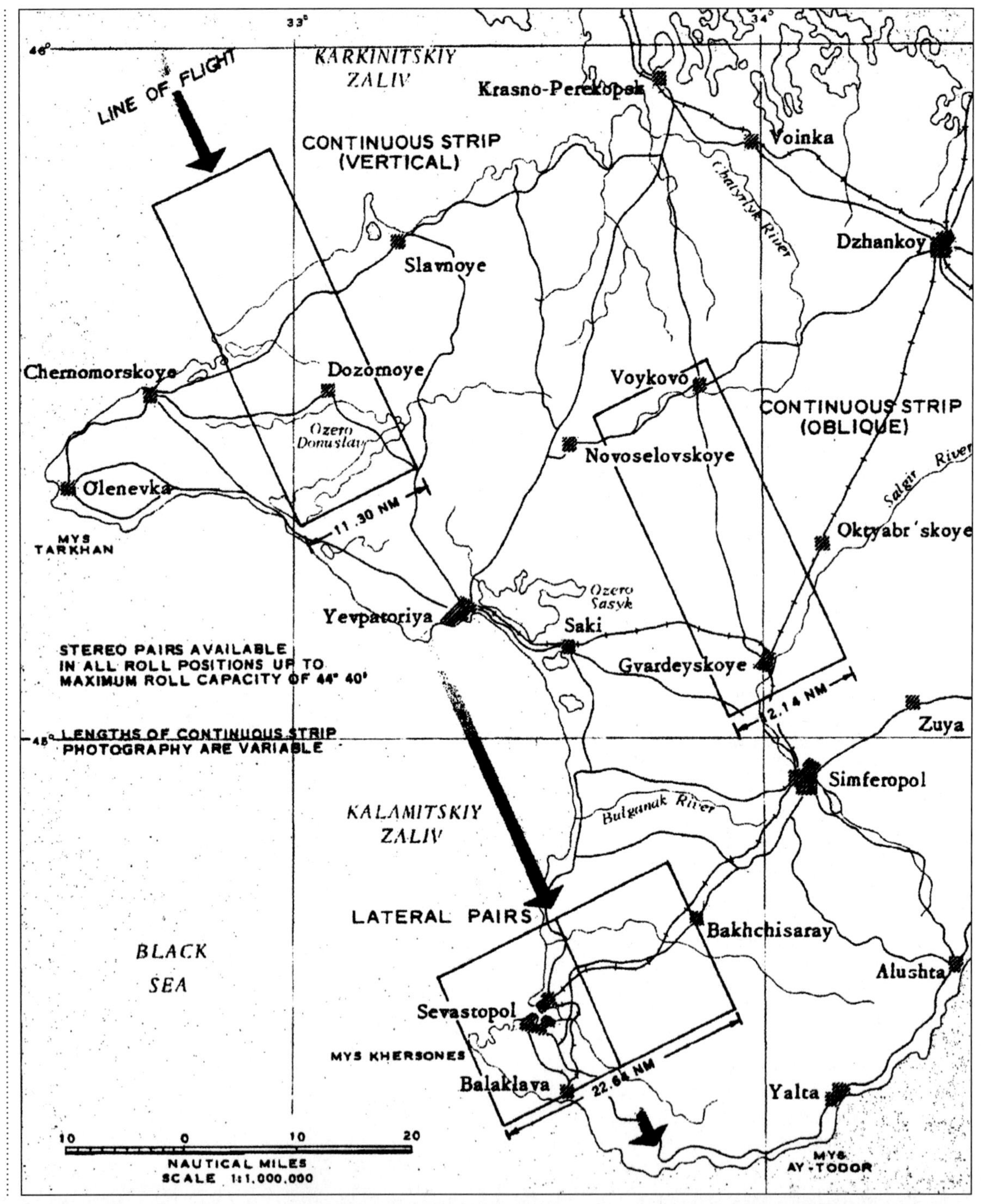

RIGHT • Examples of continuous strip and lateral pair frame coverage at a nominal altitude of 175km (109 miles). (National Reconnaissance Office)

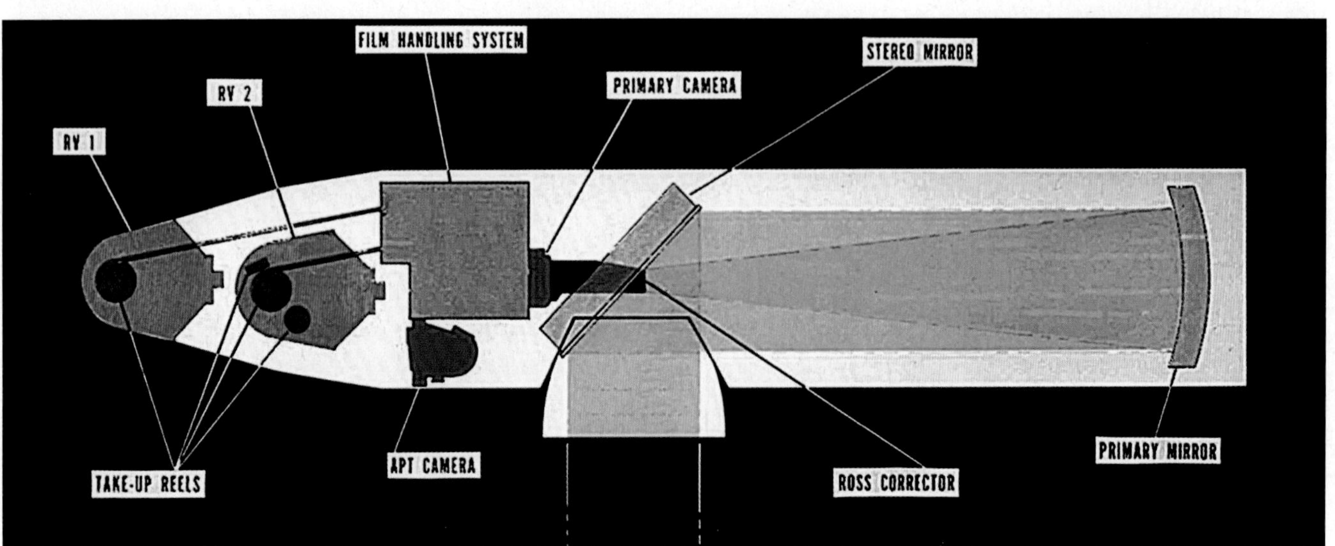

BELOW • The optical path for the GAMBIT 3 system and the film transport path to the two recovery capsules. (NRO)

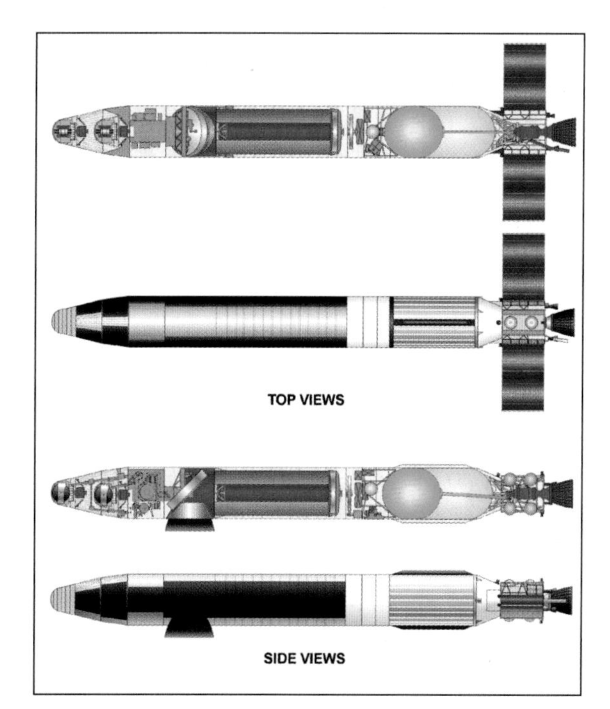

LEFT • *The Block 3 and 4 KH-8 GAMBIT 3 series with added solar arrays replacing batteries for extended duration. (Giuseppe De Chiara)*

mid-1960s, the US was operating several different intelligence-gathering satellites, including KH-4, KH-4A, KH-4B, KH-5 and KH-7. But there were technology capabilities that could take photography from space to the physical limits of optics. That limit is about 5-10cm (2-4in) from an altitude of around 160km (100 miles) and it was achieved with the KH-8/Gambit-3. For the purist, the actual designation was Gambit^3 (Gambit-cubed) but the former name was commonplace.

The primary technical challenge was with the 111.7cm (44in) diameter Kodak aspherical primary mirror and the flat stereo mirror which formed an ellipse of 147cm (58in) by 116.8cm (46in). Each mirror was larger than many astronomical telescopes but had to be much lighter. More than 800 hours of grinding and polishing was required with total manufacturing time of at least 3,000 hours on early sets. Not least challenging was the amount of work contracted by the NRO and NASA, the latter wanting a similar system for their Lunar Orbiter programme.

Some discussion was held over the balance between Gambit-1 and Gambit -3 systems and for a time, the two were flown concurrently, the first KH-8 launching on July 29, 1966. At this date, only eight KH-7s remained to be flown before retirement, some of the procurement of Gambit-1 going to the Gambit -3 line. Over time, the success of the KH-8 brought extraordinary demands on the photointerpretation and analysis teams, stretched already by the vast quantities of information flowing through the system from several concurrent satellite systems.

Great improvements were also being made to film types. From an original Kodak Type-3402 to high-definition Type-1414 and SO-217 high definition, fine-grain film, the ultimate development emerged as the SO-315 with highly uniform silver-halide crystals for highly accurate film speed and resolution. In addition, a separate line of colour and special films for near-infra-red images was developed. Much of this work took film science and manufacturing techniques to unprecedented levels of resolution and coupled with the stability of the KH-8, Gambit-3 achieved the ultimate in wet-film imagery.

During the 1960s, international events brought increasing need for time-urgent intelligence information faster than could be obtained by buckets of film. While not expecting to achieve the high resolution of film, digital images stored on board the satellite and transmitted to the ground over a suitable station was considered as a further development of the Gambit-3 system. There was a precedent for this in Samos, originally part of the Air Force WS-117L programme of 1956.

The technology was quite basic. In place of film, an array of light-sensitive diodes would convert photons collected by a large mirror and focused through a lens for conversion into an electric current. Known as electro-optical imaging (EOI), the challenge would be to make the diodes small enough so that several hundred per 2.5cm (1in) could provide resolution as good as the Corona system. Each diode would produce a single pixel to assemble the image like dots on a printed page. The digital data stream would be transmitted to the ground and in that way provide an endless quantity of imagery limited only by the electrical power supply on the satellite.

Options argued over

Samos had competed with Corona and there were 11 launches between October 11, 1960 and November 11, 1962. It had been intended to provide five optional cameras designated E-1 to E-5, with E-1 and E-2 as direct readout and E-5 and E-6 as film-return satellites. E-3 was abandoned while E-4 was for mapping functions but cancelled.

At best, the digital readout cameras provided resolution no better than 6m (20ft) while the film return types provided a resolution down to 1.5m (5ft) which, being no better than the Corona satellites obviated the need to keep Samos going.

As the Gambit-3 programme progressed, the possibility of applying digital readout to the satellites of this series was argued over at length, the conclusion eventually being reached that it was unsuited to the design. Digital transmission of reconnaissance images would only become a reality with the KH-11 Kennen, the first of which was launched in December 1976. KH-8 satellites would be sequentially upgraded and improved, taking advantage of the Titan III series of satellite launchers, adapted and developed versions of the Titan II ICBM.

BELOW • *A GAMBIT satellite image of the Capitol building in Washington DC in a deliberately downgraded resolution to veil the actual quality. (National Reconnaissance Office)*

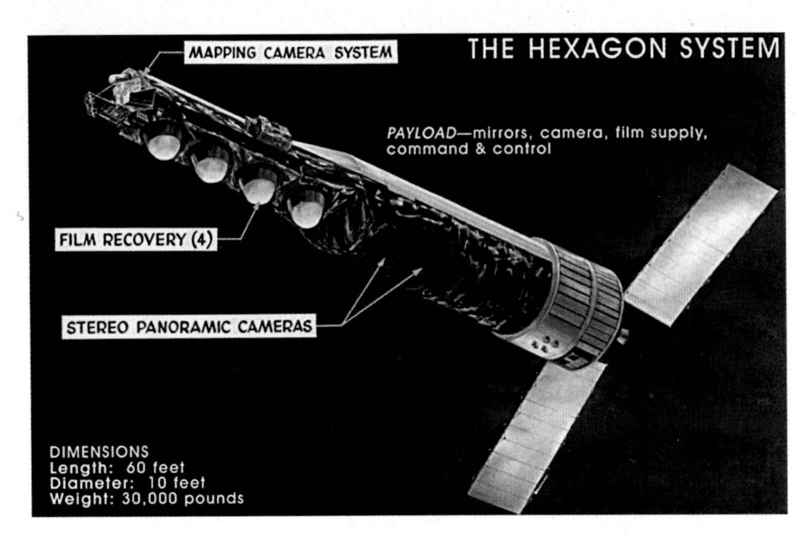

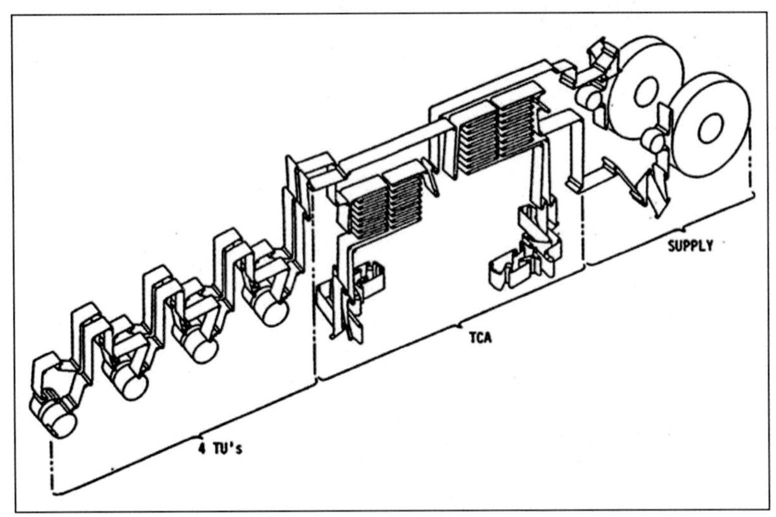

TOP • A radical departure from the Agena-based spy satellites, the KH-9 Hexagon with independent propulsion, deployable solar arrays, and a four-bucket primary film recovery system. (Giuseppe De Chiara)

ABOVE • The film transport mechanism for the two parallel delivery spools was necessarily more complex than the two-bucket satellite configuration of the Corona and GAMBIT designs, which added difficulties in maintaining film tension and temperatures. (National Reconnaissance Office)

RIGHT • The Hexagon with deployable solar arrays and a four-bucket primary film recovery system. (Giuseppe De Chiara)

The end of the KH-8 series coincided with introduction of the KH-9 Hexagon, the ultimate expression of the bucket-carrying reconnaissance system before total replacement with digital transmission satellites. But delays to the introduction of the KH-9 gave Gambit-3 a new lease on life when each mission was applied to a dual-mode application, operating from higher altitude for area surveillance over a three-month period and then down low for high-resolution targeting for which it was designed.

Intelligence gathering had reached high levels of sophistication and reconnaissance satellites were integrated with target planning aided by weather satellites and other means of selecting suitable sites to visit, several thousand on each flight. Overall observation determined that 65 per cent of the Earth's surface is normally cloud covered and against a prediction that only 35 per cent of ground targets would be visible, integration with the weather satellites allowed timely reallocation of targets so that the window of visibility increased to 80 per cent as the skill of finding cloud-free skies with desired ground objectives were balanced.

The last of 54 KH-8 G Gambit-3 flights was launched by Titan 24B on April 17, 1984, with a long list of highly credible achievements. Since the launch of Gambit-1 on July 12, 1963, a total of 92 satellites of the KH-7/KH-8 class had been launched over 21 years with a high degree of success. Their discoveries transformed intelligence gathering and revealed the true extent of Soviet and Chinese military developments, not least being the extent of production at the nuclear submarine base at Severodvinsk in the White Sea, vital to understanding the threat to merchant vessels in time of war.

The full extent of Soviet ICBM and SLBM evolution had been mapped, as well as providing detailed information on the basing sites, ground silos and a wide range of other information which was put to advantage when writing and underpinning arms control agreements. It was now possible to go to the Soviet arms control negotiators and tell them what they had, circumventing subterfuge and disinformation which had stimulated an arms race when Western countries over-estimated Soviet aircraft and missile inventories and their true level of technical development.

It was this misunderstanding that had led John F Kennedy to gain political traction over the seemingly unresponsive and languid approach by the Eisenhower administration to space 'firsts'. In the belief that the Russians posed a very real danger, US ballistic missile production and bomber inventories had grown out of all proportion to the realities of countering the Soviet war machine. That, in turn, had led the Russians to escalate their own nuclear force capacity in an attempt to catch up with the Americans. By the mid-1960s that imbalance in favour of the US and NATO was clear from the vast quantity of satellite data now streaming back to Earth stations.

Buckets galore

It was during this period, when the advantages of satellite intelligence were self-evident, that the final development of the film-return concept emerged. Since the original Corona/Discoverer series, the capability had been limited by the performance of the launch vehicle, maturing from the Thor-Agena to the Atlas-Agena and then to Titan III derivatives for Gambit-3 (KH-8). With the availability of a more powerful Titan IIID with two massive solid-propellant boosters, the payload to low Earth orbit increased to 12,300kg (27,100lb). There was no further need for an Agena stage and the satellite would be placed in orbit by the second stage of the Titan IIID. Now the full range of technical innovation was available for operational use.

What emerged from a series of development steps beginning in 1963, when discussions were held for development of a large, area-surveillance satellite, was a considerable extension in both capability and performance. Within two years, it had been agreed that the new system would have the Byeman code Hexagon and eventually it would be known as KH-9. The CIA would develop the sensors and the optics while

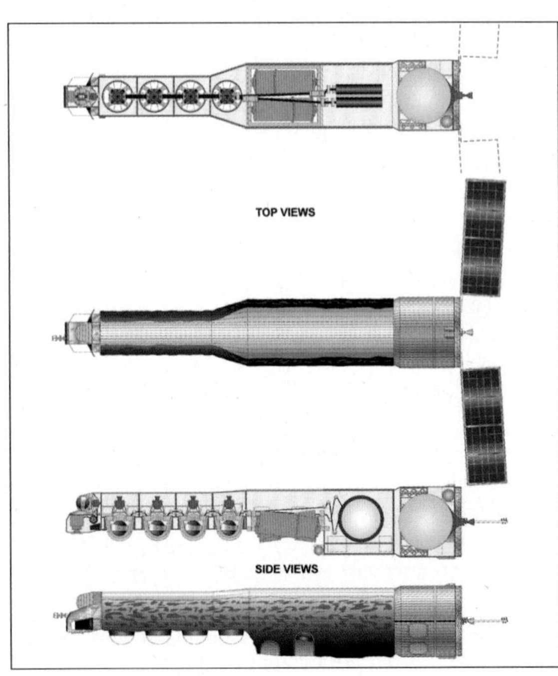

the NRO would be responsible for the satellite itself and its operational functions. Instead of two buckets, KH-9 would have four, sometimes five, and operational life would be extended to several months on orbit.

This was a major step forward and would provide a more persistent and enduring option for target planners and, due to its extended duration, it would allow inputs to changing geopolitical situations in a timely manner. This enhancement would also reduce the number of satellites built and the launch vehicles procured. Both were costly and, by acquiring fewer of each, overall funds could be reduced. Ground handling costs would remain largely the same due to the installed services, personnel, and facilities. But it was not for these reasons that the KH-9 would play such a vital role in shaping national policy. It would demonstrate new technologies for applications in other programmes, whether in optics or the design and operation of satellites.

Film from the primary search and surveillance missions would be returned to Earth in four large re-entry vehicles (RVs) based on the GE Mk 8 warhead. Film from the mapping camera would be returned in a Mk 5 re-entry vehicle. At launch, the complete satellite vehicle had a length of 17.9m (58.7ft) and a diameter of 3m (10ft) with a total mass of 12,250kg (27,000lb) but 1,270kg (2,800lb) of that was a large protective shroud jettisoned during ascent when the ascending stack exited the atmosphere. An adapter attached to the forward section of the second stage of the Titan carried a mild detonating fuse to sever a circumferential beryllium strip after reaching orbit.

The aft section was 2m (6.5ft) long and 3m (10ft) in diameter, weighed about 1,590kg (3,500lb) and consisted of an equipment module, the separation assembly, the Orbit Adjust Module (OAM) and the Reaction Control Module (RCM). This took the place of the attitude control and propulsion systems previously part of the Agena D stage on earlier satellites and carried the propulsion units and the solar arrays. The aft section was the interface between the various elements of the satellite and connected it to electrical systems, battery coolant and fluids and gases prior to launch.

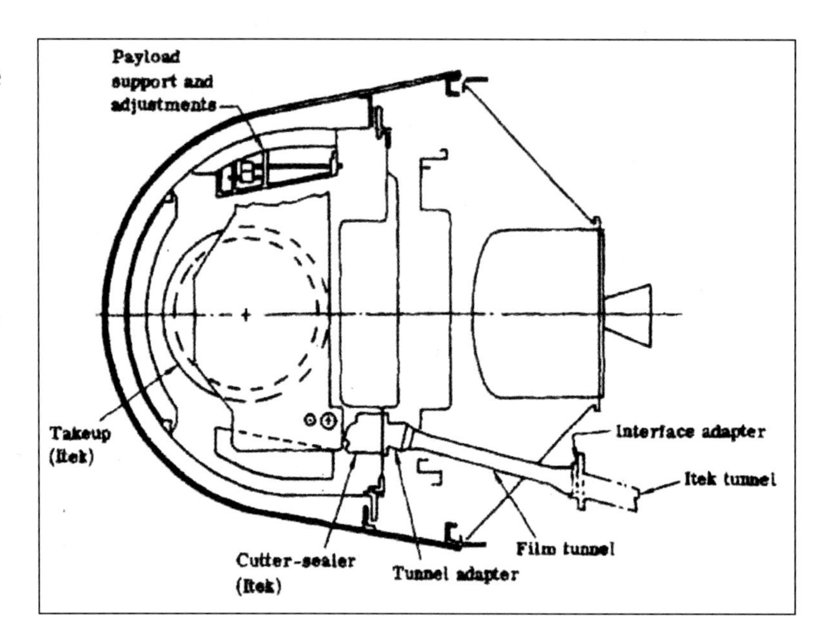

ABOVE • A simplified cutaway of the Mk V re-entry vehicle, showing film transport spool and take-up drum. (National Reconnaissance Office)

Control requirements during active use of the search and surveillance cameras on Hexagon demanded an attitude accuracy of 0.7° in pitch and roll and 0.64° in yaw, with a rate accuracy of 0.014deg/sec in pitch and yaw and 0.021deg/sec in roll. Because the attitude excursions were crucial to satisfactory operation of the primary cameras, the settling time was 0.2sec for stereo imaging and 6sec for mono photography. In non-horizontal mission mode, attitude accuracy was required to be one degree in roll and yaw and three degrees in pitch, with rates accuracy on 0.15deg/sec in all three axes.

The propulsion system used catalytic decomposition of a pressure-fed, hydrazine monopropellant, the thrust of the main engine declining from 1.112kN (250lb) to 0.445kN (100lb), with attitude control thrusters declining from 26.7N (6lb) to 8.9N (2lb). The propellant tank for the main engine would vary with the mission but typically was a spherical chamber with a diameter of 157cm (62in) loaded with 1,815kg (4,000lb) of propellant. The four 56cm (22in) diameter thruster hydrazine

LEFT • The film recovery capsules situated under the forward section, one of which has its cover removed, showing the internal arrangement. (Dwayne Day)

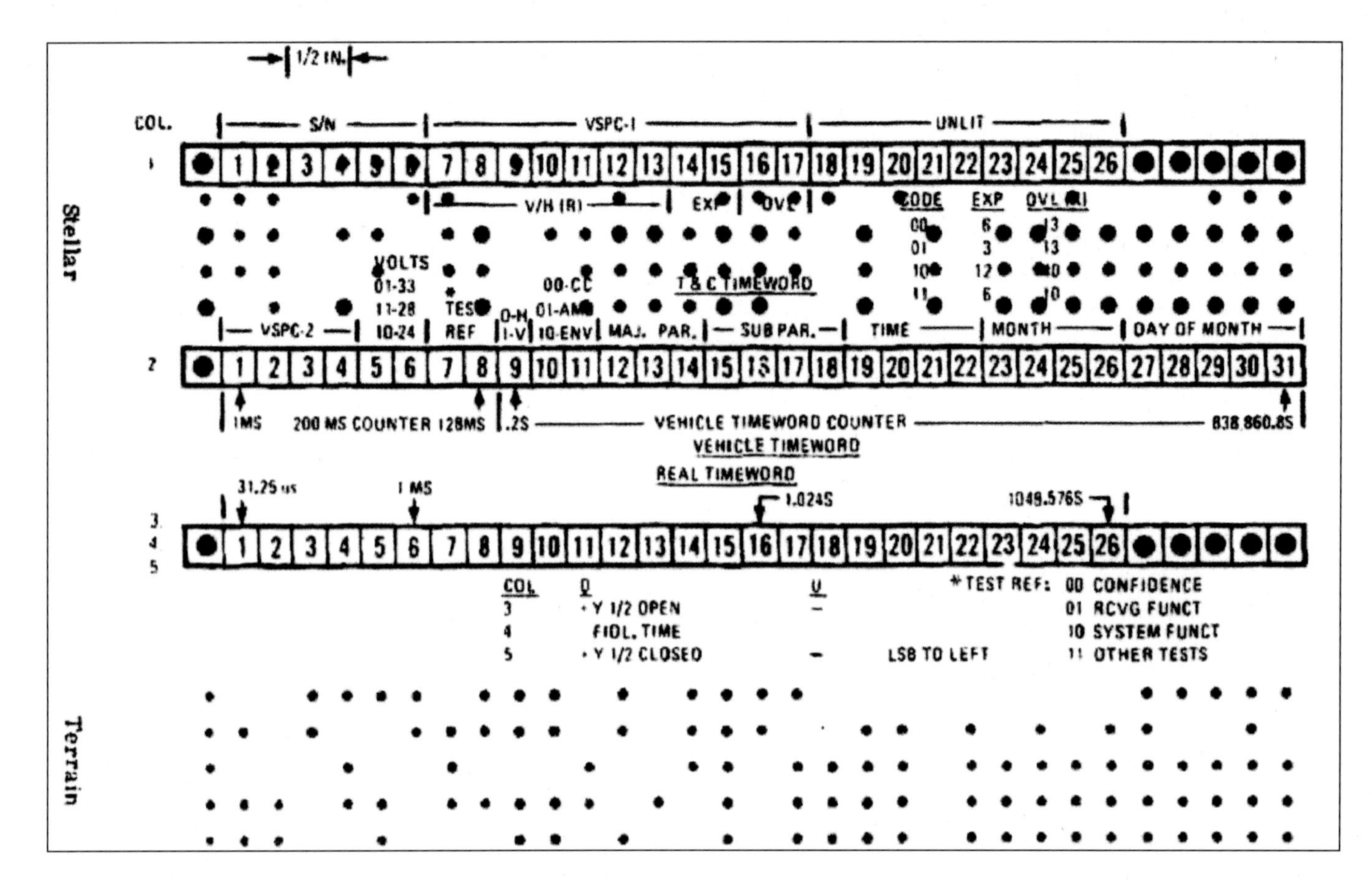

ABOVE • *Terrain camera data blocks with the readout overlay superimposed. Four projectors in the lens housing image have fiducial marks at the corners at the time of exposure. They give the precise location and the time the moving terrain image was exposed. (National Reconnaissance Office)*

tanks carried 204-249kg (450-550lb) of propellant, quantities depending on the mission. The KH-9 carried a completely separate set of eight paired thrusters for redundancy.

Primary means of maintaining power on the KH-9 comprised two folding solar array wings, each extending 5.2m (17ft) comprising 22 panels with an area of 16.44m² (177ft²) of photo-voltaic cells, one each side of the aft section and deployed on orbit. Energy storage came from rechargeable nickel-cadmium batteries with distribution at 24-33 volts dc. There were four parallel segments with an array, charger, and battery in each to minimise power collapse from a single failure and fusing was applied to critical circuits. The system could provide 11kW hrs/day of power.

The mid-section of the KH-9 had a length of 5.9m (15.4ft) and contained the stereo panoramic cameras, their optical systems, support structure, mirrors and all the associated control mechanisms for their operation, together with a film feed mechanism for transporting the exposed film to the four re-entry buckets housed on the long axis of the tapering forward section, which had a length of 6.8m

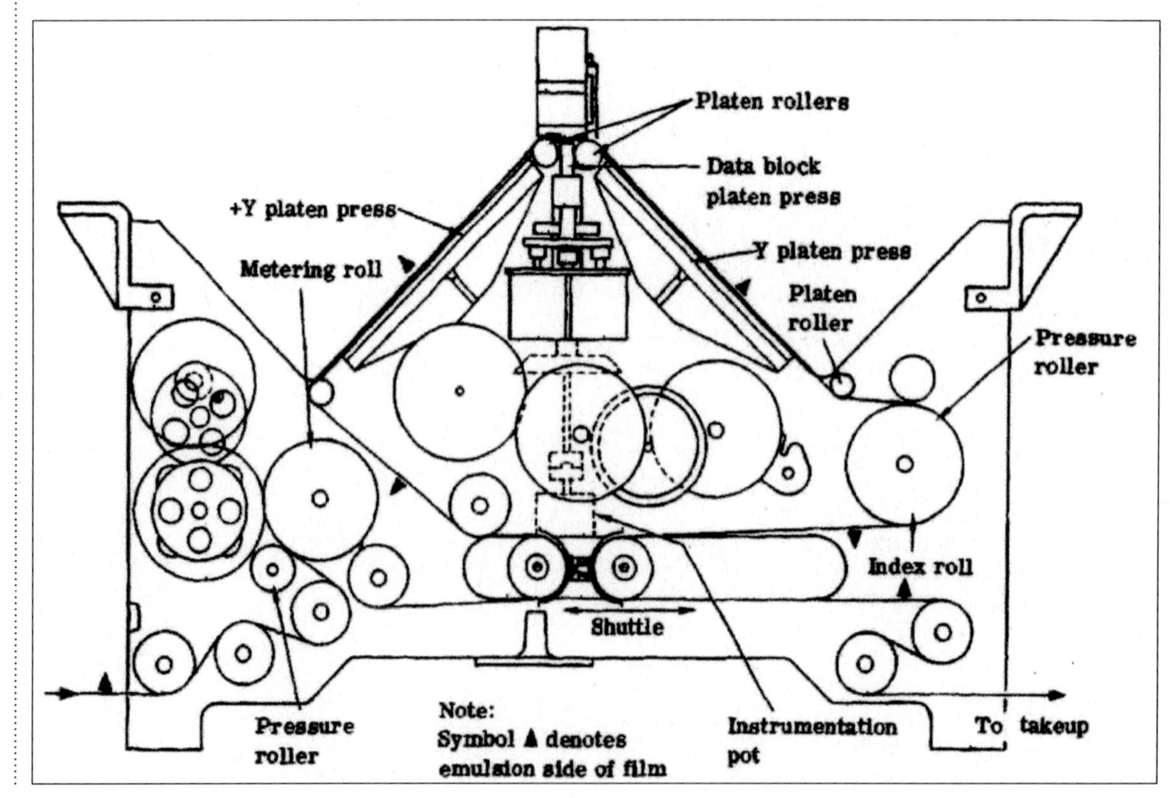

RIGHT • *The stellar film transport employed the same film advance approach as that used in the terrain film transport system, albeit with a different mechanism. The metering and index rollers were 5cm (2.5in) in diameter and with 1.72 revolutions per frame, they advanced the film 27.4cm (10.8in) with a frame interval of 7.0-80.sec. (National Reconnaissance Office)*

(22.5ft). The Mapping Camera and its dedicated film bucket were in the extreme forward end of this section. Built by Perkin-Elmer, the two separate and independent panoramic cameras were each controlled individually. They provided target resolution of at least 0.82m (2.7ft).

The optics had a 152cm (60in) Schmidt telescope which passed the image to an aspheric corrector plate and then to a 45-degree mirror. That reflected the image to a concave primary mirror with a diameter of 91cm (36in), which directed the light through an opening in the flat mirror and through a four-element lens onto a film platen. Film was of Type 1414 (SO-208) and could also be SO-130 infrared and SO-255 for natural colour at a strip width of 17.7cm (6.6in).

Unprecedented capacity

Compared to earlier satellite capacities, the amount of film carried by the KH-9 was unprecedented. Varying according to mission requirements, it was typically 47,245m (155,000ft) for each camera, which was about sixteen times the capacity of that in the GAMBIT satellites, and a total weight of 907kg (2,000lb)!

The two film supply drums were offset in alignment with each other across the width of the mid-section and staggered vertically. These aligned respectively with the forward-viewing camera on the port side and the aft looking camera on the starboard side. Multiple loop reels carried the film under tension to the twin-camera assembly further along in the mid-section and on to individual take-up spools for each bucket. The controlling electronics were situated in the lower part of the mid-section as viewed with the long axis in the direction of orbital travel.

The mapping camera and its associated systems were attached as a separate unit to the front of the satellite and contained its own re-entry bucket. In all, 20 Hexagon mapping cameras were built and 12 were flown, beginning with the fifth flight of the KH-9 on 9 March 1973. Mapping operations covered 171.7million km^2 (50million miles^2) of denied territory and global coverage of the rest of the world at a rate of 34.3million km^2 (10million miles^2) per year. This was for the preparation of maps with sufficient accuracy to provide land, sea, and air forces with accurate reference to carry out combat operations or targeting at any point on the Earth's surface.

Hexagon flies

The first flight of the KH-9 (vehicle 1201) occurred on June 15, 1971. It remained operational until August 6, followed by the second on January 20, 1972, a year which saw three flights of the KH-9. Hexagon flight rates numbered three in 1973, two in 1974, two in 1975 and one in each of the next five years. There were no KH-9 flights in 1981 but one in each of the following three years. The first KH-9 to fly on the still more powerful Titan 34D was launched on June 20, 1983, a useful switch as the overall weight of the KH-9 increased with various modifications and improvements to film loads.

Throughout the programme, piggy-back electronic ferret satellites flew on 1203, 1207, 1208, 1209, and on the eight flights from 1212 to 1219, boosted to higher orbits where they tracked Soviet radars, recorded missile tests, and gathered radio communications among other tasks. Flights 1210 and 1212 also carried science sub-satellites released on orbit to conduct their own clandestine operations, a not uncommon pairing. Doppler beacons had been carried on 1205-1207 for measuring the density of the atmosphere at high altitudes.

The last launch of 20 KH-9s took place on April 18, 1984. It was the only flight to fail when 1220 was lost in a catastrophic accident shortly after lift-off, an inglorious end to a magnificent record of unprecedented success. While the resolution of the mapping camera had at first been only 9m (30ft), it improved with development through the programme and by the end interpreters were getting photographs with a resolution of 6m (20ft). Outright resolution was not the goal, the 29,000 photographs returned to Earth exceeding the goal when the KH-9 programme was approved.

The KH-9 was the ultimate bucket air-recovery system, incorporating high resolution area surveillance and wide area mapping and the system was unique in achieving that. Formed on January 1, 1972, the Defense Mapping Agency was almost exclusively reliant on Hexagon. Missions 1201-1204 provided main camera imaging, while missions 1205-1216 provided mapping camera products, the last three successful flights reverting to the area surveillance function.

With the US Air Force committed to flying a large number of military payloads on the NASA Shuttle, in January 1972 studies looked into the reuse of Hexagon satellites – either by refilling them with propellant in space, conducting on-orbit maintenance, or capturing them in the cargo bay for return to Earth, refurbishment and relaunch. The Shuttle had only just that month been approved.

As it turned out, the Shuttle did not fly before April 1981 but the following year a number of astronauts were cleared for briefings on the Hexagon and NASA was involved in examining the possibility of using the Shuttle to send up the last two KH-9 vehicles on this reusable launcher. The costs were too high, and the idea was dropped, the logic of going that route with a system on its way to retirement being problematical.

BELOW • ***A GAMBIT** image of the Sary Shagan Soviet science research facility and home to the suspected development of directed-energy (laser and particle-beam) weapons. (National Reconnaissance Office)*

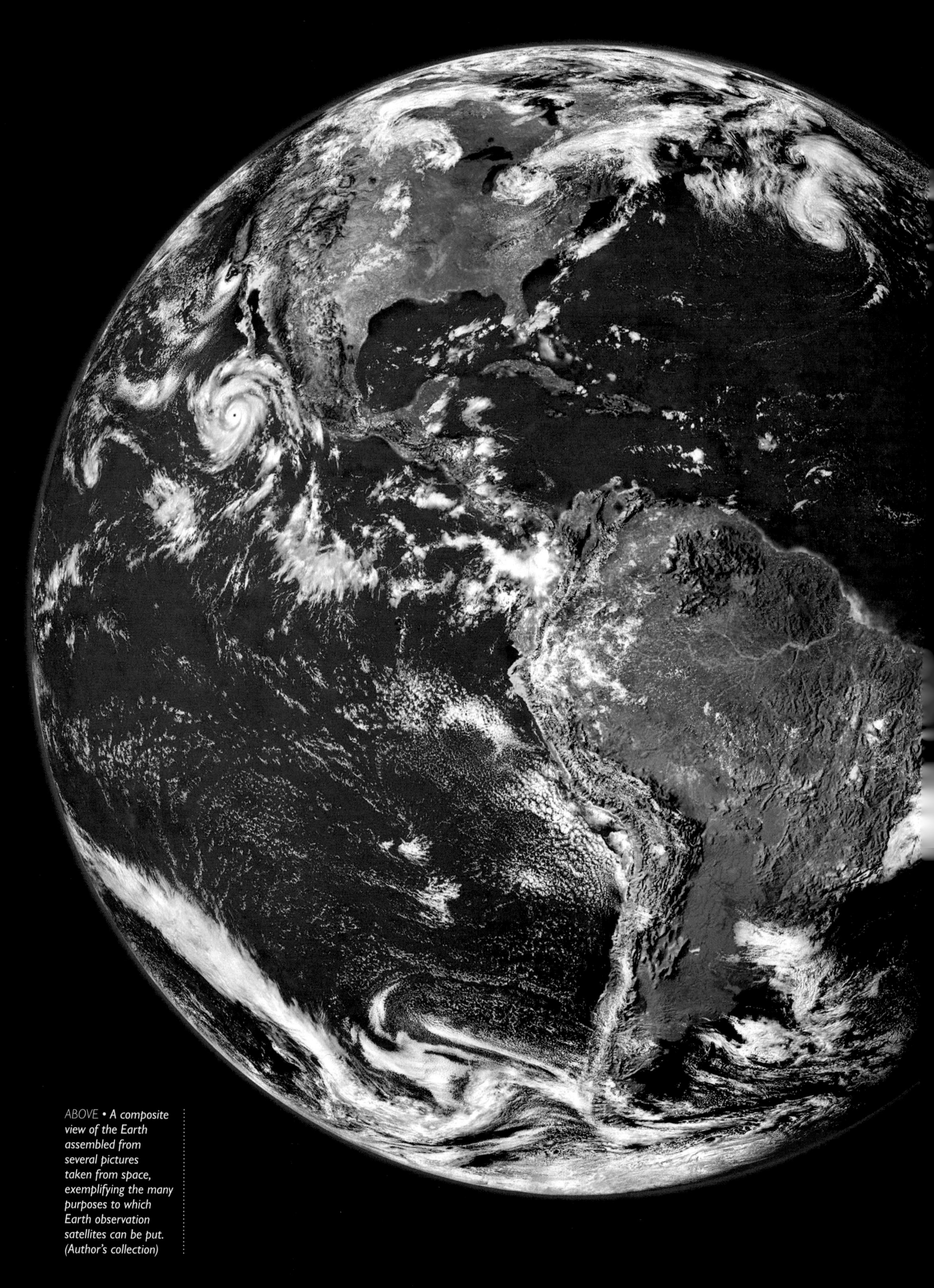

ABOVE • *A composite view of the Earth assembled from several pictures taken from space, exemplifying the many purposes to which Earth observation satellites can be put. (Author's collection)*

DIGITAL EARTHWATCH

The economic use for space technology started becoming clear.

From the early 1960s, the development of reconnaissance satellites for military and national security purposes opened a range of new technologies and a series of unique capabilities. The basic science underpinning all Earth observation technologies depends on passive remote sensing, where optical instruments such as a camera collect sunlight reflected from objects on the surface of the Earth. For the most part, this is in visible light, which human eyes have evolved to capture and which the brain interprets as a picture or a moving image.

What we see through our eyes is only one small part of the total spectrum, from x-rays and gamma-rays at one end to extremely low frequency radio waves at the other. In the middle lies the visible spectrum, so named because our eyes are tuned to interpret those signals as observable light. But instruments can be made to observe different parts of the spectrum, providing additional pieces of information that we cannot see. Those other parts are just as useful as reflected visible light and they can add to our understanding of what we are looking at.

To measure different parts of the electro-magnetic spectrum, instruments can be tuned to observe different wavelengths – the highest numbers being at the radio end of the spectrum and the lowest at the gamma-ray end. Reciprocally, the frequency, measured in Hertz (Hz) and named after the German physicist Heinrich Hertz, has the lowest numbers at the radio end and the highest at the gamma ray end. The frequency is measured in cycles per second, or cps where 1cps = 1Hz.

The wavelength is the distance between identical peaks in the wave, while the frequency is the rate per second of a vibrating wave. These properties determine what the wave is – be it photons of visible light, gamma rays or radio signals – and it is these units that define energy levels right across the electro-magnetic spectrum. In short, the longer the wavelength, the lower the frequency and the energy; the shorter the wavelength the higher the frequency and energy levels. Wavelength is measured in metres and energy in electron volts.

Observing the same part of the Earth's surface at different wavelengths can provide information the human eye cannot see. This is known as multi-spectral imaging. Military reconnaissance satellites were the first to exploit this from space, optical systems capable of translating into visible colours different parts of the spectrum, usually in infra-red wavelengths. It revealed information the human eye cannot see, such as the demarcation lines between vegetation and different types of mineral composition on landscapes.

The military were particularly interested in these distinctions because reflected sunlight can produce a confusing picture of what is actually being looked at. The camera can lie and give a false view if the pictures it produces are only seen in the visible portion of the spectrum. Sensors tuned to observe in the infra-red and a variation using separate frequencies can provide thermal images. These measure the energy level of an object as a function of its brightness, or emissivity, and because they are outside the spectrum of visible light, they reveal details even where there is no sunlight. Once again, by observing the wavelength, the energy level is known.

Heat means information

A good example of this is the use of thermal imaging to measure the amount of heat coming through the walls of a tent or machinery covered by a tarpaulin. Whether an object is hot or cold can tell when it was last used. Military reconnaissance satellites look for signs of heat conducted through such coverings to give an idea as to whether the covered machine has been used recently and is still cooling off.

This capability has stimulated development of non-conductive covers for military tents, vehicles, machinery, and small power generators to mask recent or continuing activity and confuse analysts by producing a 'cold' reflection. A wide range of such products are now marketed through arms traders and commercial suppliers. It has even been used for infra-red detection of illegal substances grown under heat.

ABOVE • Aerial reconnaissance and high-altitude photography has underpinned many a military victory, including this image taken of Utah Beach immediately before the Normandy landings on June 6, 1944. (Author's collection)

Thermal imaging can also reveal biological activity and filters in the near-infra-red can spot heat sources. Police units use thermal imaging cameras on helicopters to track individuals running from arrest or to follow vehicles in a chase, directing law enforcement officers in pursuing vehicles. In less dramatic form, thermal imaging over large areas can, if filtered into the appropriate frequencies, provide information as to the health of crops. Diseased crops emit higher heat levels from accelerated biological activity, readily detectable through thermal imagers.

These activities are known as passive remote sensing, splitting reflected light into component parts. Active sensing involves sending a signal to the ground and analysing that when it is reflected back to the detector. Examples include tracking of ships, aircraft, and other moving objects in the air, at sea or on the ground. Laser and radio altimeter scans can accurately measure sea levels, plot the rising bulges of water when submarines pass below the surface or map features on the sea floor. They can also be used to gather information on wind speed and direction and on the movement of ocean currents by measuring wave height, frequency, and direction of flow.

RIGHT • A photo-analyst uses high magnification lenses to interpret photo-imaging during the Korean War of 1951-1953. (USAF)

Another application is LIDAR – 'light detection and ranging', where distance is measured by firing a laser beam to a reflective object and measuring the time taken for it to return. This has several military and civilian applications. If sustained as a continuous set of pulses, it can measure the changing distance to a target of a precision-targeted missile in flight, enhancing its accuracy. Or it can be used to provide highly accurate terrain maps for geologists, environmentalists, and rural planners – and, in many cases, to provide 3D maps of ancient archaeological sites.

What can be done with multi-spectral imaging is determined by the spectral band used for observation. While there is some crossover between the various colour bands, blue is generally used for images in the atmosphere or in deep water where it is effective to a depth of approximately 50m (150ft). Green is useful for vegetation and to depths of 50m (150ft) in water. The red band is useful for soil and general vegetation, while the infra-red bands are appropriate for various types of soil, vegetation, moisture levels, distinguishing different geological features and for mapping different surface materials such as clays and silicates.

Special tuning

Between the distinct bands, and especially in the infra-red, separate frequencies can be tuned within the sensors and filters to discriminate specific types of subject observed during mapping. This is especially useful for aerial surveys of different minerals in surface features and also for different types of vegetation. The degree of moisture in the surface is also an important consideration when selecting frequencies

LEFT • Satellite imagery is used by military forces to assess the results of a strike by land or air forces, in this picture of the Podgorica Airfield, Montenegro during Operation Allied Force in the Balkans, 1999. (USAF)

for the sensors. Sometimes it is helpful to use a complex integration right across the frequencies and this is known as hyperspectral imaging, where individual pixels are scanned for content. This can be useful for identifying materials where the type is unknown, as each will respond differently when observed and thus disclose its unique signature.

To avoid excessive use of building materials and rare minerals in construction projects, thermal and hyperspectral imaging of large areas are essential for providing information to planning authorities. This type of work emerged not long after the end of the Second World War, where major developments with cameras, filters and bespoke film for aerial photography provided a base for applying ostensibly military technology to civil and commercial applications.

The technology had been around long before that. Before the Second World War, Kodak had developed multispectral colour photography and 'false' colour infrared films were used widely during the 1940s. False colour photography is a means of making visible those parts

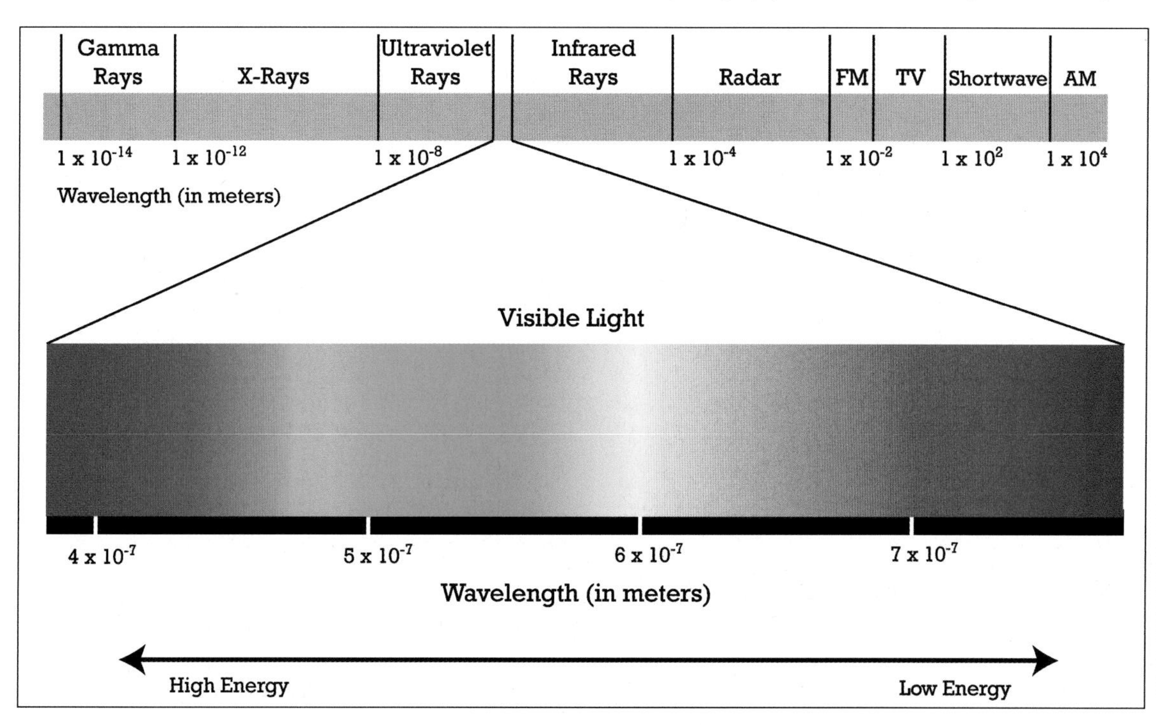

LEFT • The electromagnetic spectrum provides a wide range of windows through which to view the Earth, the shortest wavelengths having the highest energy. (Author's collection)

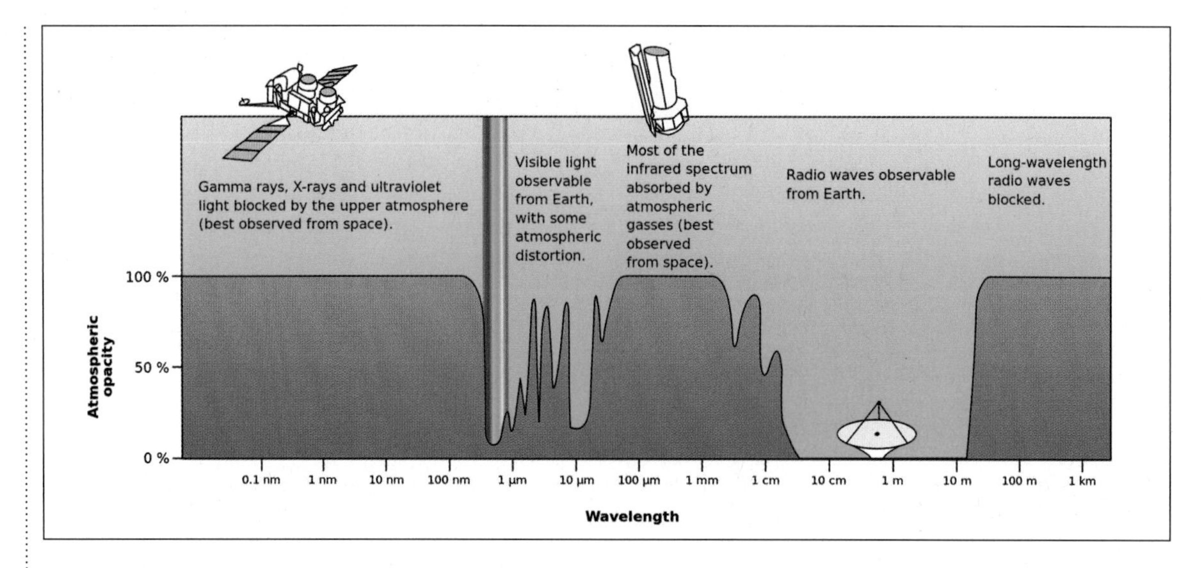

RIGHT • Visible and radar wavelengths penetrate the Earth's atmosphere to the surface and these bands are utilised by satellites. Only portions of the infrared reach the ground. (Author's collection)

of an image which are outside the range of wavelengths the human eye can see. Film usually consists of separate layers of emulsion sensitive to blue, green, and red and these are the natural colours the observer sees. To 'see' infrared images, the green emulsion is printed in blue, the red in green and near-infrared in red. In a standard colour photograph, healthy vegetation appears green but in colour infrared images it appears as bright red because it reflects light more strongly in near-infra-red light.

As early as the 1930s, laboratory experiments had shown how spectrometers could differentiate the spectral properties of plant leaves and show how different pigments in botanical samples could distinguish stepped spectral signatures in the low visible to the high near-infrared wavelengths. Through into the 1960s, further research was carried out by the US Department of Agriculture at its research facility in Weslaco, Texas, where the optical studies of leaves became a speciality. This was tested by instruments flown aboard aircraft.

The ability to obtain information from remote sensing evolved through aerial photography but the potential for space applications was not immediately apparent. Nobody really knew just how well a satellite could harvest this data, largely because the atmosphere is opaque to some sensors and optical devices operating in selected wavelengths. It was clear that aerial photography had advantages over surveys at ground level, not least due to the larger swathes of surface area that could be imaged from higher up. But there were many unanswered questions about space-based instruments. Only some of the answers had been provided by early experience with military reconnaissance satellites.

Seeking sensors

The reconnaissance satellite programmes were highly classified and there was no feed across to the civilian sector, but the fundamental technology that made those possible spilled across into the optics industry. During the first half of the 1960s, great progress was made with satellite design and engineering and the space environment was better understood. It became apparent that the concept of a remote sensing satellite was feasible and could bring tremendous advantages. Not only for the United States but for the way in which the US government could draw in many countries around the world in a co-operative endeavour to better understand the planet and work with the natural world on both urban and rural development.

The ideas of many scientists and remote-sensing specialists were pooled and discussed at length in Earth science conferences during the early years of the space programme. With interests extending to the surface

BELOW • The atmosphere viewed against the limb of the Earth's surface from space showing the depth of the atmosphere which, in terms of life support, is comparatively about the thickness of the skin on an apple. (NASA)

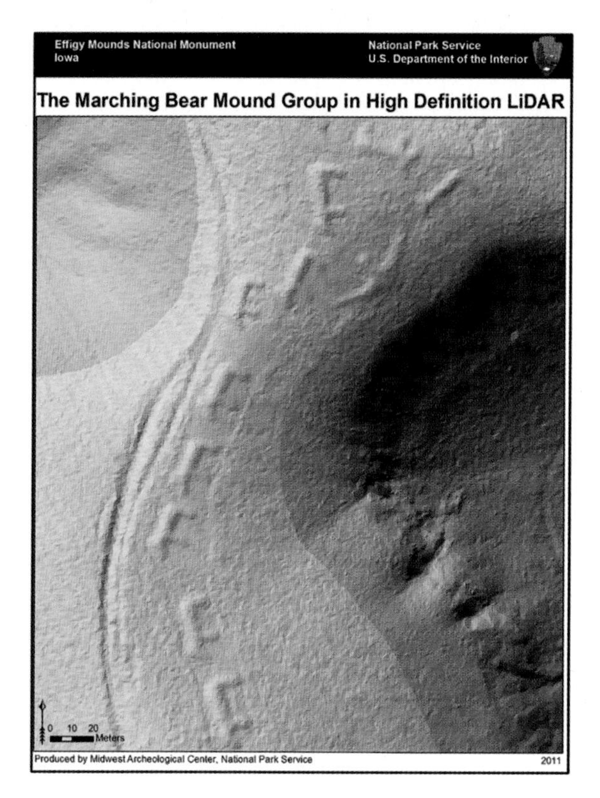

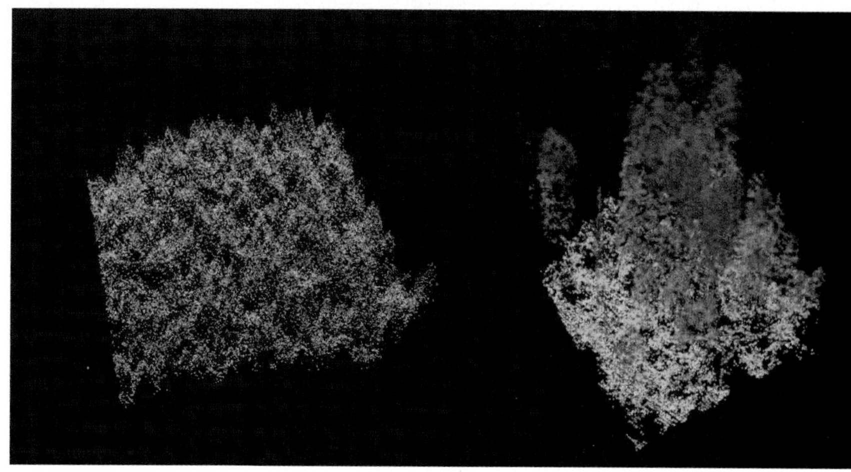

geology of other worlds, there was growing attention to the evolution of the Earth and several academic research organisations set up studies into life elsewhere – astrobiology. The use of satellites for better mapping and identification of different conditions across the Earth became a part of that, using our own planet as an analogue for biodiversity elsewhere in the solar system.

The US State Department saw a political opportunity to project a peaceful image for the United States by making such information about the planet, obtained from American satellites, directly available to countries around the world. If encouraged to build their own ground stations for receiving remote-sensing images, these countries could receive data directly from space. But it was for the value in embracing an international endeavour that many scientists put their name to the prospect of a remote-sensing satellite.

A key figure throughout, in 1965 the eminent American geologist William Pecora became the director of the US Geological Survey at the behest of President Lyndon Johnson. Pecora immediately put his position behind the drive to get a global remote-sensing programme. The idea was to provide an experimental satellite to find out just how effective remote sensing from space could be and how well a satellite could provide useful data. It could also provide research into digital transmission direct to ground stations, a technology only then being introduced to the most advanced military and national security satellites.

But there was strong opposition. The Bureau of the Budget thought it to be a waste of money – surely aerial surveys were sufficient – and the intelligence community thought it would provide foreign despots with easy access to detailed imagery for belligerent purposes. Unconstrained distribution of Earth images, they said, could undermine all the work being done to secure classified information about other countries that only the United States had access to.

A compromise was needed and as the State Department and related agencies began to back the idea, it was decided to limit the optics of any civilianised remote-sensing programme to a resolution of no better than 60m (197ft). This, it was said by the Pentagon, would not provide sufficient resolution to assist military operations. Instead, said the State Department, it would draw uncommitted countries into the mainstream global order by providing useful information for peaceful activities. This, they believed, would help stabilise dysfunctional regimes and ensure a better way of providing aid so that countries could help themselves become better custodians of their own future without relying on foreign assistance.

Economic use

With these objectives in mind, State Department officials, public service broadcasters such as the Voice of America and local consular staff in foreign countries could demonstrate how the use of information from remote sensing would expand their economies and improve their agricultural productivity.

Instructional materials were provided for schools which could request a visit from local American officials to describe how best to use the information which was freely provided.

In the United States, NASA's initial reluctance to get involved was more to do with extreme pressure on the annual budget and the diversification of resources required. But when the Department of the Interior decided that it would go ahead with a remote-sensing programme, NASA changed its mind and backed the initiative. For NASA, this was a seminal moment. Almost nobody at the space agency had anything to do with the military spy satellite programmes, although the technology was well understood. It was merely a matter of physics and optical engineering.

ABOVE • *Light detection and ranging (LIDAR) images of old forest (right) compared to the signature of a new plantation. (Sarah Frey)*

ABOVE LEFT • *LIDAR imaging of archaeological sites helps provide 3D images of semi-buried remains and is a key partner with satellite images in interpreting such places. Here, the Marching Bears Mound Group in Iowa. (National Parks Service)*

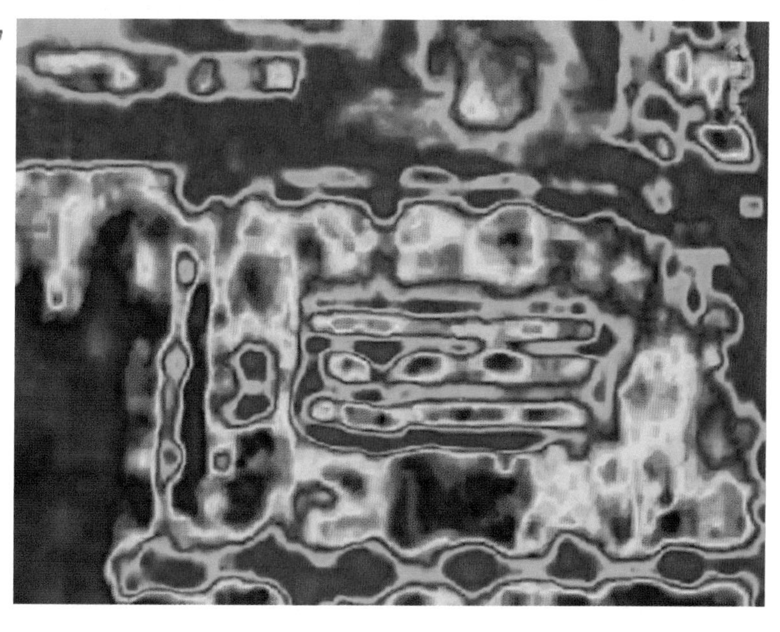

BELOW • *A thermal image of a former brickworks in Armadale, West Lothian taken as part of an archaeological thermographic map. (John Wells)*

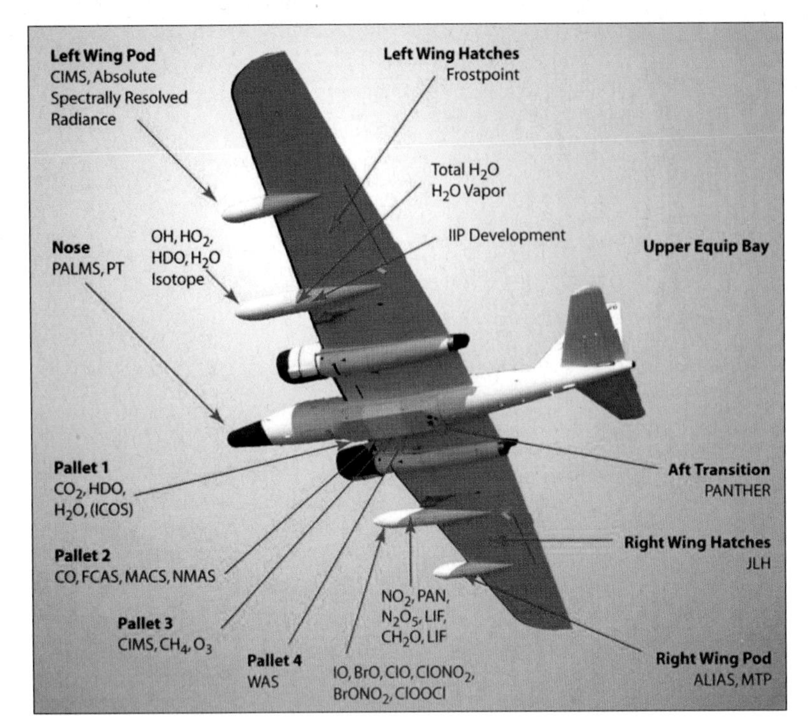

ABOVE • *NASA operates the WB-57 here shown with a typical suite of science instruments which helped develop the equipment and the roles for remote-sensing satellites. (Author's collection)*

But 1965 and 1966 were seminal years in which NASA flew 10 two-man Gemini missions on which Earth imaging with hand-held cameras and a selection of spectral filters were evaluated. Previous to this, there had been only four, short-duration Mercury missions with very little opportunity for extended photographic surveys. By the end of 1966 it was apparent that an extraordinary amount of detail and multispectral data could be obtained from a remote-sensing satellite. This made the prospect of a successful service all the more attractive.

Technical capabilities had kept pace with aspirations, as with the introduction of Itek Corporation's 9-lens camera to capture data in nine separate parts of the spectrum on the same image. New lead-sulphide and mercury cadmium telluride materials were key to getting images at longer wavelengths into the thermal infra-red, allowing scientists to deliver more accurate interpretation of land materials and surface conditions. These developments went along with demonstrations by camera-carrying aircraft collecting data from a single spot on the ground where a scanning mirror was placed in front of the sensor. Combining the accumulated lines produced an image.

A major breakthrough in providing multi-spectral imaging was made at the University of Michigan when an optical-mechanical system was taken into the air on test, a development of the original M5 two-channel system designed to sample infra-red wavelengths only. In 1963 two aircraft flew across the same test area, each imaging at slightly different wavelengths. When overlaid, the four channels provided what is generally regarded as the first multi-spectral imaging with more than two channels. The challenge then was to build a four-channel system into one instrument.

Progress was rapid. Within three years the university was testing an 18-channel system, none in visible light but since it was constrained to aerial photography the various channels were not registered with the scanner data and so could not claim true multi-spectral capability. By 1971, the M7 scanner had begun trials imaging 12 bands and into the visible portion, thus allowing calibration.

This work had a long history at the university, whose scientists did much to persuade dissenting opinion that remote sensing for satellites might have more application in civilian sectors than the military. As early as 1962, the university's Infra-red Laboratory at its Institute for Science and Technology hosted the first symposium on the new instruments, attracting 70 delegates. In 1968, the eighth symposium had more than 700 attendees.

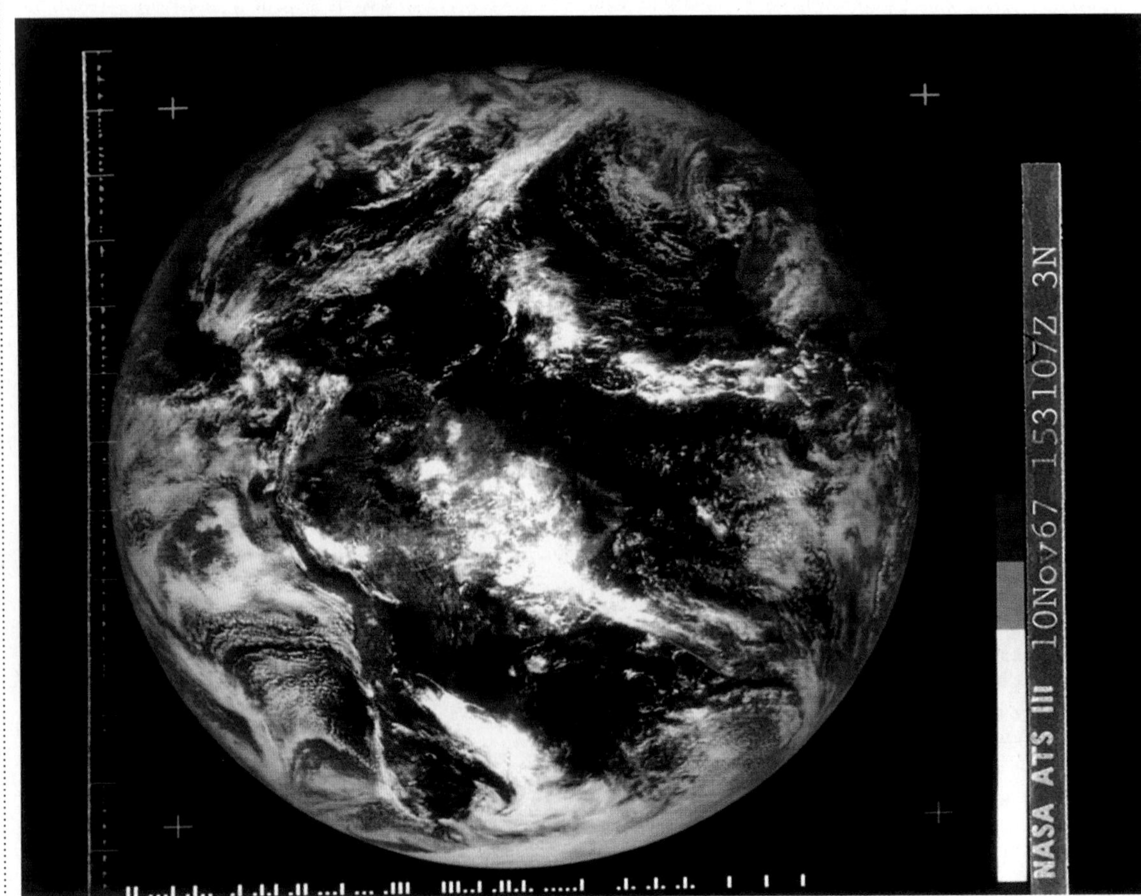

RIGHT • *Launched in 1967, NASA's ATS-3 satellite transmitted the first digital images of the whole Earth as a portent of remote-sensing and Earth science satellites. (NASA)*

BELOW • *The massive KH-11 was launched into orbit by Delta IV-Heavy with the first going into space on December 19, 1976. (NRO)*

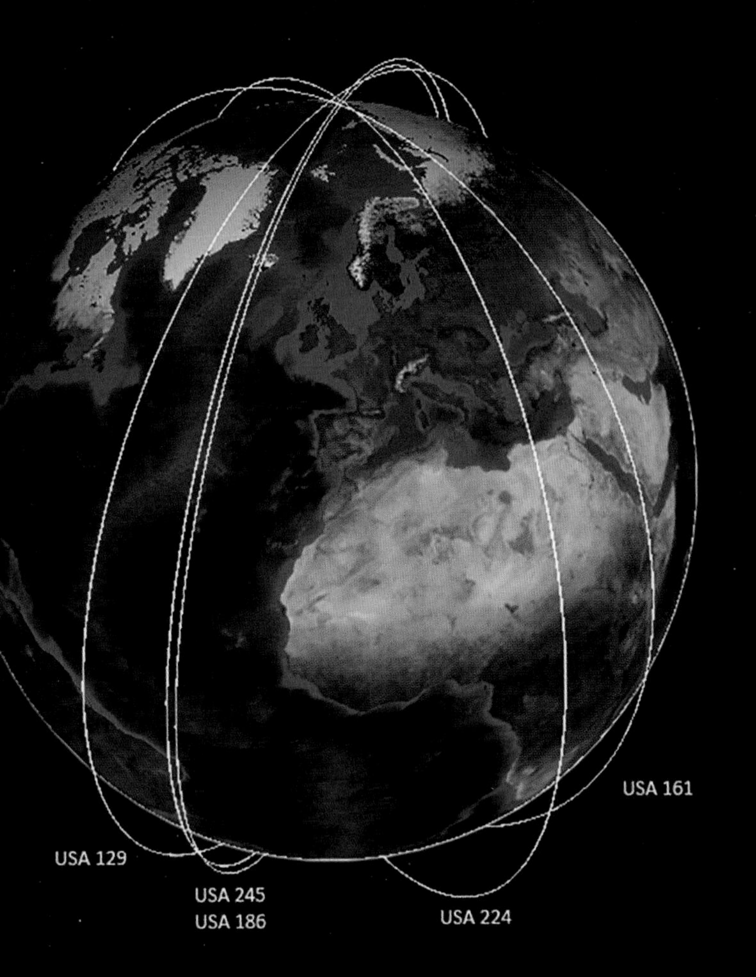

ABOVE • *KH-11 satellites operate as an integrated constellation to maintain a near-constant digital surveillance over denied territory and across various regions of interest. (NRO)*

Growing awareness

All this was supported by a growing awareness in the scientific world that unconstrained use of limited natural resources was disproportionately changing the Earth. That a global catalogue of the planet's environment was essential to better understanding of the human impact on the natural world. The burgeoning use of highly advanced and sophisticated technology, from runaway industrialisation to nuclear energy and atomic weapons, stimulated a growing concern among sections of the general public over unintended consequences for the habitability of the planet. That itself did a great deal to push forward plans for a remote-sensing satellite and it got the attention of relevant government agencies.

As NASA itself began to look beyond the then upcoming Moon landings, in 1966 it provided money to help start what would eventually become the Laboratory for Applications of Remote Sensing (LARS) at Purdue University, Indiana, which quickly forged a working relationship with the University of Michigan. Involved from the outset were the US Geological Survey (USGS) and the US Department of Agriculture (USDA), both enthusiastic about development of these technologies.

The work at LARS-Purdue developed the first innovative system for digital processing and image analysis and digital computers. This spilled across into several satellite applications involving data handling and interpretation, initially with the M5 and M7 camera tapes. Sceptical at first, scientists and engineers there developed an efficient optical-mechanical multi-spectral scanner and at a meeting of the National Academy of Sciences in 1968 produced a set of results particularly relevant to agriculture. It was this that would form the basis for the satellite's multi-spectral scanner.

Landsat

The urgency with which advanced multi-spectral remote sensing was required became self-evident with the sudden emergence in 1969 of corn blight across southern Texas. Previously known only in tropical zones, it was affecting increasingly large areas of the Midwest and Southeast. The USDA worked with the universities of Michigan and Purdue to support NASA as it flew specially modified WB-57 aircraft with colour infra-red cameras along 6,400km (4,000 miles) of flight lines in 400 hours. The information gathered was crucial to protecting future crops.

Instead of an unacceptable use of pesticides, which would have been uneconomical over such large areas, 25 per cent of farmers switched to blight-resistant corn on information which was shown to be much more effective than field observation and ground-based data-gathering. This alone spurred support from farmers and agriculturalists for a multi-spectral eye in space, one constantly monitoring the fields. Analysis indicated that limiting operating conditions in aerial observation with sunlight reflection and atmospheric interference would not be constraints in satellite observations.

As noted previously, NASA had not been enthusiastic about the central role it would be required to play in any remote-sensing satellite programme alongside post-Apollo plans for grand expeditions involving Moon bases and manned landings on Mars. But support for these ambitious goals from the public and Congress was waning and to increasing numbers of the population, NASA they seemed less relevant. With Civil Rights issues and an escalating war in Southeast Asia, the space agency was being challenged on its priorities.

By the late 1960s Congress was unsure about the future role of NASA, questioning whether it had any purpose after winning the Space Race with its impending Moon landing. Many politicians believed it had only one central purpose and that once it had demonstrated the reason it had been formed – to achieve pre-eminence for the United States – its role was redundant. In 1968 even presidential candidate Richard Nixon wondered whether it could not be repurposed into a general technology development agency and would ask that question again after reaching the White House and cancelling the last two Moon missions.

Seeking to rebalance its national priorities, as early as 1965 NASA had convened a conference on the use of satellites and spacecraft for geographical research and for aiding with land management by using satellites to conduct an inventory of natural resources. This gathered support in Congress and a level of internal momentum at NASA until, by the end of the decade, the agency was beginning to openly diversify into a broader range of space applications.

This contrasted to some extent with the way other government departments saw the future for Earth observation. While supporting a broader diversification, NASA also wanted to preserve a role for humans in space. Desperately protective of its manned flight programme, anticipating the end of Apollo Moon missions in the early 1970s it sought a series of manned laboratories, one function of which would be to manage Earth surveys from space. Other government agencies sought the use of unmanned satellites, cheaper and more frequently flown to incorporate increasingly more capable equipment.

Ironically, the CIA and the National Reconnaissance Office had been fighting the same battle with the Air Force, which wanted manned spy satellites in orbit for reconnaissance and military surveillance. In its drive to put uniformed officers in space, the Air Force had developed the Dyna-Soar spaceplane but that had been cancelled in 1963 and replaced with the Manned Orbiting Laboratory (MOL). This was a space station, the studies for which dated back to a strategic space plan of 1959 for long range Air Force objectives, and which had an anticipated initial launch in the early 1970s.

As it emerged, MOL would have used an Air Force Gemini B to ferry astronauts back home and most configurations would have transmitted images to Earth. MOL was a highly ambitious programme and while it would have begun with a cylindrical module supporting a Gemini on the front end, more advanced versions were considered. These envisaged a large complex of multiple modules, observation platforms and a crew of up to 40 people operating telescopes, multi-spectral imaging instruments and optical cameras operating in the visible bands.

However, during the 1960s, the intelligence community took advantage of strides in digital transmission and in the development of infra-red imaging. That resulted in the development of the KH-11 Kennen (perception by deduction) satellite which would grow over the programme to have a weight of around 19,500kg (43,000lb) and provide digital, electro-optical images. The first KH-11 would be launched on December 10, 1976 and successive generations would continue to the present but that is a different story. Throughout this period, while the intelligence community was operating the early KH-11 types along with the KH-9 Hexagon, the first civilian remote-sensing satellite programme got under way.

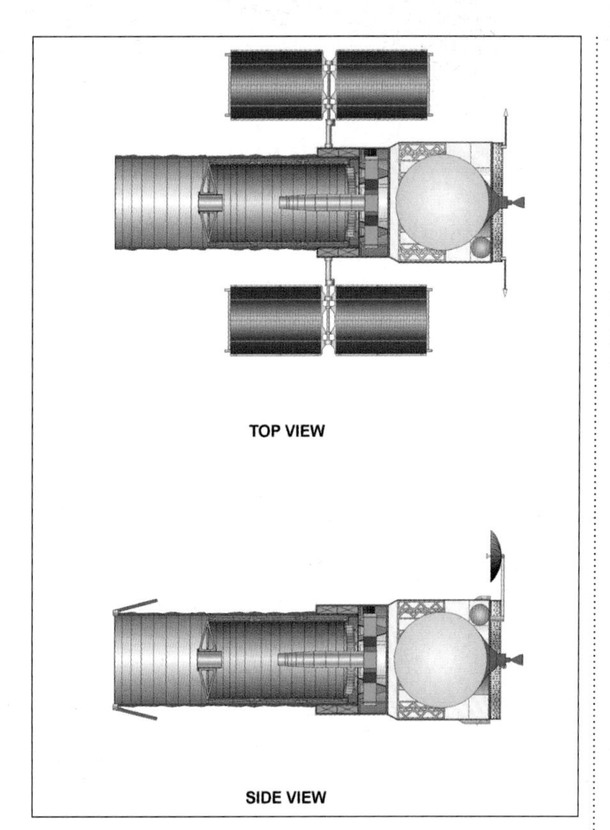

LEFT • The 1970s saw the introduction of the KH-11 Kennan digital transmission reconnaissance and area surveillance satellite operated by the National Reconnaissance Office, the template for which was adopted by NASA for the Hubble Space Telescope. (Giuseppe De Chiara)

Renewed examination

With self-evident logic for a non-military Earth observation system, NASA began a renewed examination of options between manned and unmanned systems and concluded that it would not have the resources to support large orbital facilities for Earth science activity. Better instead to develop systems for cheaper and more frequent access to space and begin a supporting programme of diverse space applications which in the vanguard would be an Earth science initiative.

BELOW • The Zhawar Kili Al-Badr camp in Afghanistan as viewed by KH-11 in support of allied military operations. (NRO)

Much of this would be managed out of the Goddard Space Flight Center (GSFC), which over time grew its involvement within the environmental and Earth studies community.

But the community of Earth scientists became frustrated with indecision on the part of NASA and the US Department of the Interior (DOI) moved ahead with its own plans for an Earth Resources Observation Satellite (EROS). Release on September 21, 1966, the news came as no real surprise to NASA, but it had the endorsement of the USGS and its boss, the pioneering William Pecora. Udall was concerned about indiscriminate use of the environment and mobilised a national effort among his equals to gather data and bring that to politicians. And he had influence in high places, having had the announcement released while he was rafting down the Colorado River with the president's wife – 'Lady Bird' Johnson.

Furious at apparently being upstaged over its vacillation, NASA sought to take charge of any remote sensing satellite programme and had the National Academy of Sciences convene with a workshop of scientists in late 1966, where it formally announced its own proposal. This was to be known as the Earth Resource Technology Satellite (ERTS) programme, with an anticipated launch in 1972. President Johnson disregarded the EROS concept and backed ERTS, the clunky acronym eventually replaced by the more descriptive name Landsat, which we will now use here.

Getting it ready

Major decisions had to be made regarding the type of satellite required and the specific instruments it would carry. Demanding requirements placed priority on a stable orbit and a satellite that could maintain highly accurate pointing angles for its instruments. More than a decade into the Space Age, there were three candidate platforms from which the Landsat satellite could be developed.

There was the Tiros weather satellite, a research initiative developing toward an operational system for global meteorological coverage in a co-operation programme with the National Oceanic and Atmospheric Administration (NOAA). Originally part of a joint venture with the US Weather Bureau, the Nimbus satellite programme was the precursor to a fully operational weather satellite. And NASA's existing Applications Technology Satellite (ATS) which would soon evolve into NOAA's Geostationary Operational Environmental Satellite (GOES), a multi-agency programme for storm tracking and general research into meteorological phenomena.

BELOW • *The chief designer on the LANDSAT Multi Spectral Scanner (MSS) system, Virginia Norwood is known as the 'mother of Landsat'. (NASA)*

The ATS programme was developing the potential for geostationary satellites, where their orbital period equals that of the Earth's rotation, and they appear to remain stationary over a specific place on the surface. Satellites in this position located in the same orbital plane as the equator have the advantage of viewing the same side of the Earth on a continuous and unchanging basis. That location is ideal for viewing changes over the same area of the Earth. Or, for sending relayed radio or TV signals to countries directly in view of a geostationary satellite.

But that is not the type of orbit where observation of the whole Earth is possible. For that, a polar orbit is ideal, where the satellite moves in a near circular path around the Earth at 90° to the equator, passing over the two poles on each revolution. A typical low-Earth orbit lasts 90min. As the Earth rotates in one full revolution of 24 hours, the satellite will pass over the same ground track after 16 orbits. Each day, the satellite will 'see' the same strip of the planet as it did 24 hours before. It is this orbit that provides the best viewing platform.

However, that is not the best orbit for viewing the same places with the same lighting angle. As the Earth moves around the Sun every 365.25 days, the lighting angle of a true polar orbit will gradually change over time. Since multi-spectral imaging ideally benefits from the same lighting angle on repetitive visits a true polar orbit skews the reflected light over the course of a year over any given spot. To allow the orbit of the satellite to process through a complete revolution each year, the satellite will revisit the same sites with the same light angle.

This is known as a Sun-synchronous orbit (SSO) and is achieved by placing the satellite in a near-circular path of around 600km (373 miles) by 800km (497 miles) inclined at about 98° to the equator. The precession is caused by the Earth's equatorial bulge which causes the orbital plane to move around in a cone-shaped path, much like a spinning top slowly 'wobbles' around its vertical position.

It was with these demanding constraints and requirements that the Landsat programme went looking for contractors to provide the satellite. The main structure supporting electrical power, attitude control, attachments for the instruments, a command-and-control system for communications, a downlink capability and thermal control was known as the 'bus'. The instruments and the sensors were the 'payload' on the bus.

Following a deep analysis of available satellite buses, the General Electric Nimbus satellite was closest in overall design and specification and the company was awarded a contract to tweak the design and apply it to a line of Landsat satellites. The payload would consist of a Return Beam Vidicon (RBV) camera developed by the Radio Corporation of America (RCA) and a Multi Spectral Scanner System (MSS) from the Hughes Aircraft Company.

Landsat mother

Known as the 'Mother of Landsat', Virginia Norwood was chief design engineer on the MSS and this equipment was available within nine months of contract. Initial testing took place by the end of 1970 at the Yosemite National Park. Built for the first five Landsat satellites, the MSS was equipped with a 23cm (9.1in) fused silica mirror bonded to a triple-bar system in the base of a nickel/gold frame with a secondary mirror which merely had to oscillate around the prime optical axis to achieve focus.

NASA's Goddard Space Flight Center led the programme, and the image processing software was developed and managed by Valerie L Thomas who also set up and ran the Large Area Crop Inventory Experiment (LACIE). It was through the LACIE tests that global crop monitoring was demonstrated and proven to be an invaluable part of the benefits from Landsat, her work influencing a worldwide application of this throughout the Landsat programme to this day.

The RBV camera was regarded as the prime instrument and would operate like a conventional photographic camera, shuttered and with the capacity to freeze an image. It was regarded by many potential users as a successor to the traditional film camera and, in several respects, it was the forerunner of today's digital cameras. It was also believed to be more reliable than the proposed MSS, offering users a natural progression from the aerial cameras with which they had been working. In several respects it was a sophisticated TV camera with 34 times the resolution of an analogue television of the time. It would provide images with a resolution of 80m (262ft) in size.

The RBV was in fact three cameras operating in the visible and near-infra-red bands with blue-green, orange-red and the near-infrared spectral regions. NASA had reservations about using the RBVs, its engineers questioning the image quality. There were concerns too about the technology being used and to ensure reliability RCA applied the lessons learned from the development of cameras for the Ranger Moon missions, which transmitted images immediately prior to impact. But the MSS was not without its problems too as several redesigned units preceded the final configuration.

BELOW • *At NASA's Goddard Space Flight Center during the 1970s Valerie L Thomas developed the software for the digital image processors. (NASA)*

DIGITAL DOWNLOADS

ABOVE • NASA's third Applications Technology Satellite (ATS-3) captured this whole-Earth view of the planet in 1967, expanding the possibilities of space-based data. (NASA)

BELOW • Secretary of the Interior from 1961 to 1969, Stewart Udall was instrumental in pressing for remote-sensing satellites and is seen showing Lady Bird Johnson, President Johnson's wife, the spectacular scenery of the Colorado River. (Robert K Knudsen)

Optimism is high as the first images from space are eagerly awaited.

Anticipation was high as launch day for the first Landsat mission approached, with 700 scientists from around the world having applied for funds to develop research projects based on its data, 200 from the US and 100 from overseas finally approved. It had truly international potential and optimism was high as users around the globe eagerly awaited the first images.

In a supportive mood that brought new laws for environmental protection, the US government signed off on the Clean Air Act and the Environmental Protection Agency was formed. In 1972, Congress passed laws for the protection of marine life and brought in the Clean Water Act, imposing harsh penalties for contraventions. Already on the books was an Endangered Species Act which would become law in 1973. For its part, the United Nations set up standards in environmental protection required of member states. Even as preparations for the launch of Landsat 1 were completed, scientists began to wonder whether the lauded satellite was really up to the task!

Built by General Electric at its Valley Forge, Pennsylvania, facility, the Landsat 1 satellite had a circular, drum-shaped base with a diameter of 1.5m (4.9ft) supporting a truss structure at the top, on which were mounted satellite systems for communications. With a weight of 953kg (2,101lb), it had a total height of 3m (9.8ft). The RBV and MSS optical equipment was mounted inside the base drum and two large solar array panels were attached to the top of the truss structure providing electrical power.

Two tape recorders were provided each with 550m (1,500ft) of 0.78cm (0.3in) re-writeable magnetic tape holding 30min of data for a total storage capacity of 3.75GB of data. That in itself was a historic achievement, providing the greatest capacity of any satellite recorder system launched thus far. It was cleared for up to 1,000 hours of operation, sufficient for the predicted life of the satellite.

In a celebrated event watched by programme personnel, scientists and engineers working the mission, Landsat 1 was launched by a Delta 900 on July 23, 1972. The launch vehicle was a derivative of the Thor missile with an extended-length main stage and nine solid propellant rocket boosters, jettisoned on the way up. Launch took place from Vandenberg Air Force Base. As with the polar-orbiting Corona and Gambit military reconnaissance satellites, launching due south from the coast of California avoids rockets flying over land.

It took several days to get Landsat 1 operational but technical troubles hit the satellite from the beginning. Although operating well, the RBV cameras had to be turned off for the duration when an electrical circuit caused problems. But the MSS worked as anticipated and proved a success. At an inclination of 99.2°, the orbit of the satellite was adjusted periodically so that it would return to the same part of the Earth's surface every 18 days. From an average altitude of 947km (588 miles), the MSS was able to produce useable images of a swath of the surface 100km (62 miles) wide.

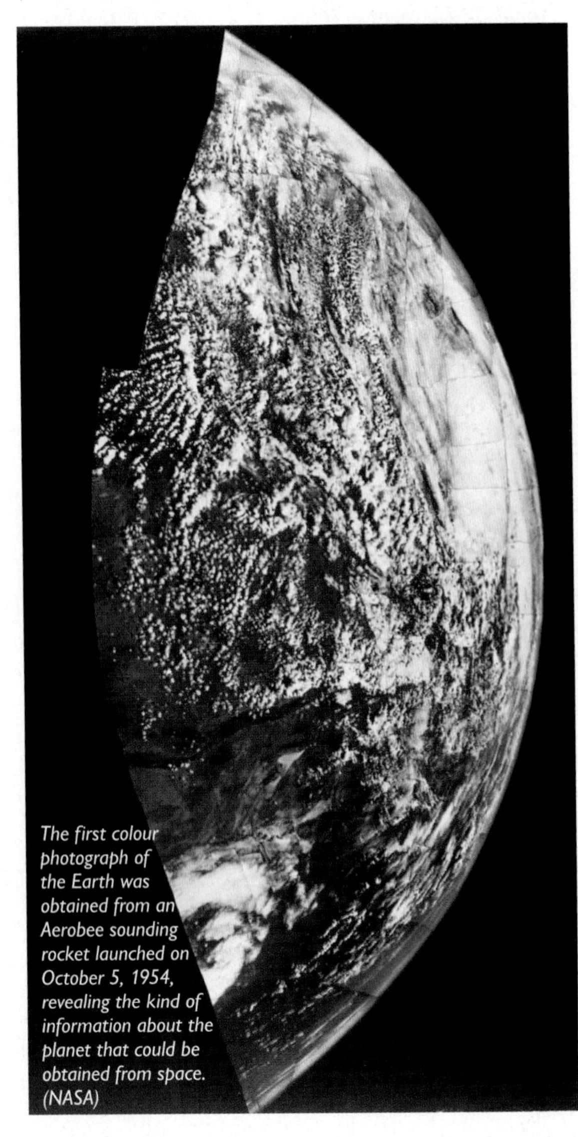

The first colour photograph of the Earth was obtained from an Aerobee sounding rocket launched on October 5, 1954, revealing the kind of information about the planet that could be obtained from space. (NASA)

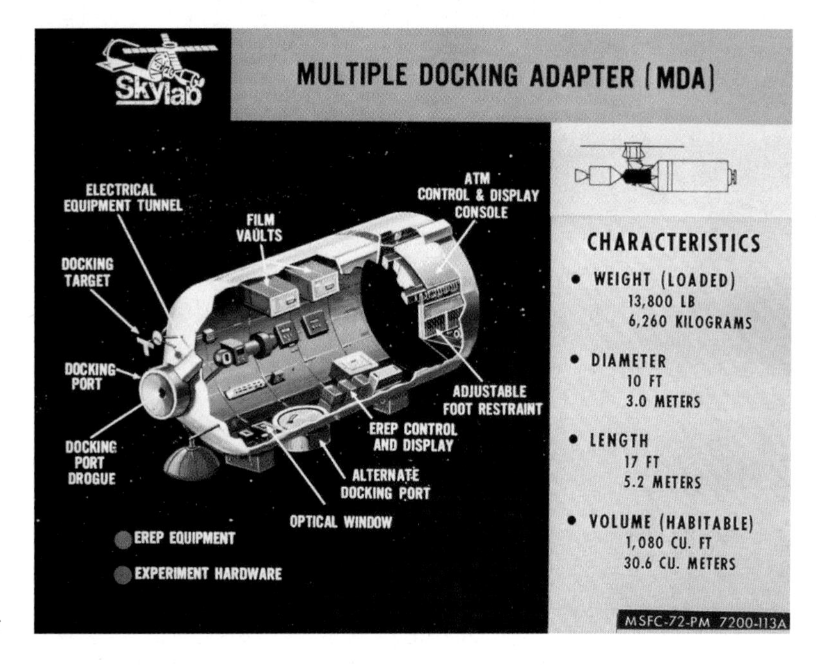

ABOVE • The manned Skylab space station operated in 1973-74 supported a module which carried cameras and sensors in its Earth Resources Experiments Package (EREP). (NASA)

Gargantuan task

Collecting and using the data from Landsat was a gargantuan task and a major challenge. Given the global nature of the satellite's coverage, many places were put in their exact geographical locations for the first time. Since the first sailing ships discovered new lands around the globe, physical maps were only as accurate as the navigational skills of the ships' crews. Even maps of known and populated areas differed according to the national authorities concerned. For Landsat, a global reference system was essential if data was to be shared and used by several different countries.

To provide what became known as the Worldwide Reference System (WRS), the US Geological Survey introduced a universal grid covering the entire planet, allowing identification of a specific site. For instance, Washington DC became WRS-1, Path 16, Row 33. It would take several years but the overlay on Landsat images to an oblique map projection was key to successfully aligning specific places with identifiable locations. The work was completed by John Parr Snyder, a mathematical cartographer who successfully put intermediate transformational equations on to a Mercator projection of a spherical Earth.

It took a lot of work and several years but from this, any images could be projected on to a common map projection. That opened the possibility of extremely accurate geolocation exploited later for satellite navigation systems. There is a direct linear development path between military reconnaissance satellites, remote sensing from space and GPS (Global Positioning System) route planning. The opportunity for these applications, and later for the World Wide Web, would come from technologies developed by the military. Exploited by commercial markets, widespread public use would create the network of satellite-based electronic and digital infrastructure we all use today.

Few of these applications could have been predicted 50 years ago with the advent of the Landsat remote-sensing satellites. Nor could it be foreseen that the technology coupled to clear advantages with operating such a system would grow a global industry of remote-sensing programmes operated by numerous countries around the world. Management of Landsat, however, was key to its success and NASA's Goddard Space Flight Center was responsible for daily operations from its facility in Greenbelt, Maryland.

Operated around the clock, it pioneered sustained control of satellites dedicated to Earth science and the monitoring of the environment, weather, and climate. Initially during the period of experiment and trial, downlink stations were operational in Alaska, Maryland, and California, to which on-board Landsat data recorders transmitted pictures. Test operations focused at first on the principal investigators, giving them priority targets with which they could judge its effectiveness, moving to optimised locations as the weather permitted.

It was in this operation that the interaction between viewing options and weather conditions provided valuable real-time planning to maximise the opportunities provided by the satellite's position and the value of the images recorded. This had been the experience when matching the viewing opportunities with spy satellites to local weather conditions. With Landsat, it helped to have a direct transmission capability, relaying directly to the ground on an S-band signal.

Identical satellites

The initial plan funded two identical Landsat satellites, each with a life of one year – and despite uncertainty as to the design life the expectation was that the first would last a lot longer than planned. The first satellite had been a great success, returning more than 100,000 images in its first three years and discovering a new island off the coast of Canada, named Landsat Island. Landsat 1 would continue to send data from the MSS instrument until January 1978, when its tape recorders were no longer operating. The satellite was shut down on January 6, 1978, when its orbital precession placed it in continuous sunlight, causing it to overheat.

Launched on January 22, 1975, by Delta 2910, Landsat 2 entered the same orbit as Landsat 1 but with a nine-day offset, which halved the interval between WRS-1 ground tracks. It was essentially the same as Landsat 1 and outlived its predecessor, lasting seven years before it was shut down on February 28, 1983, due to attitude-control problems.

While highly successful in their planned objectives, Landsat's 1 and 2 were crucial in helping to avoid financial loss in the farm sector and to discount propaganda put about by the Soviet Union regarding their crop yields. While the Air Force, and to a lesser degree the Army, had wanted accurate maps of military targets, land features, terrain conditions and natural obstacles from their spy satellites, the CIA sought information about societal conditions in the Soviet Union. This included agriculture and the true state of the Soviet economy, the reality behind the myths and the propaganda.

In Imperialist Russia, the country had been a net exporter of grain, with most European countries buying up supplies

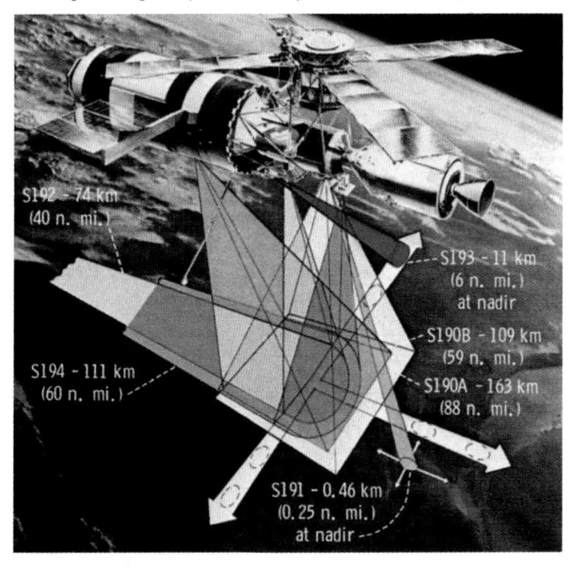

LEFT • The Skylab EREP instruments provided the first manned operation of Earth-resource analysis at a time when the Landsat programme was becoming operational. This image displays the footprints for various EREP instruments. (NASA)

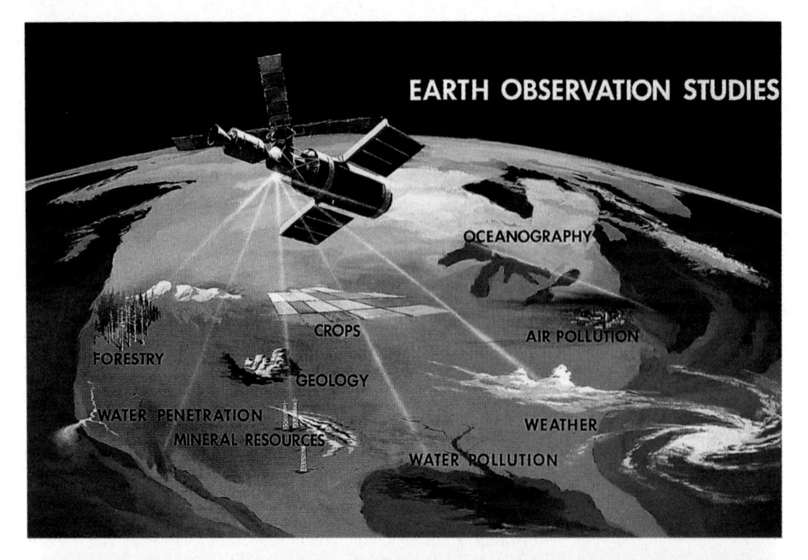

TOP • The 1970s saw a major increase in studies of the Earth and its environment, the warming of an industrialised planet and the desire to use space for obtaining the data later generations would use in attempts to mitigate climate change. (NAS)

ABOVE • Connecting remote-sensing satellites to ground stations required a relay system provided by the Tracking and Data Relay Satellite System (TDRSS), a model of which is displayed at the Steven F Udvar-Hazy Center in Chantilly, Virginia. (Baton Greyjoy)

to keep their growing populations fed. After the Bolshevik revolution of 1917, the push for moving Russia's agrarian population into collective farms caused widespread starvation and death, forcing the Soviet Union to become a net importer of grain. With demand increasing over time, by 1972 Russia was buying up 25 per cent of the entire US wheat harvest, leading that year to a massive price hike to prevent depletion of domestic grain silos and shortages in the US.

The US government had no means of verifying Soviet propaganda about its food production inventories and the Landsat programme was alerted by the UN to conduct a global catalogue of agricultural production. This resulted in a major survey, a trigger for the LACIE programme mentioned in the previous chapter. It was through this programme that back-channels were set up to supply Russia with information which would help moderate its predicted purchase requirement and prevent an imbalance in supply and demand from America.

The State Department declined to make political advantage of this information. To the outside world, it was a closed issue, but it ensured stability. Moreover, the Nixon administration was keen not to offend the Soviet Union as it sought its influence in bringing the North Vietnamese to the negotiation table. In a wider context, it was through accurate crop yield prediction that world food production came into balance with annual requirements on the global market and introduced a significant increase in output. This was not through more intensive farming but through better management and enhanced yield projections, a major triumph for the Landsat programme.

But there was a lot more to come. Landsat 3 was based on the first two satellites but with significant improvements under the skin. The RBV cameras were modified into a twin-camera system with an improved spatial resolution of 40m (131ft), twice the value for the first two, and operating in one broad spectral band of green to near-infrared. The MSS was a collective set of spare hardware and had an additional thermal imaging band, although one failed early. The last of the 'new' parts for the MSS contributed to a cascade of problems during flight.

Launched on March 5, 1978, by a Delta 2910, Landsat 3 was timely in its arrival on orbit as it was assigned to extend the global survey begun by the first two satellites. The orbit maintained the nine-day repetition with Landsat 1 and as operations got underway the MSS began to falter and would never work completely as expected for the remainder of the satellite's operating life. The RBV cameras too were troublesome and never really achieved expected performance, although the images were unaffected. The satellite was placed on standby mode on March 31, 1983 and decommissioned on September 7.

Growing the potential

Times were changing and the world was different. Twenty years into the Space Age, the achievements that pushed robots to the planets and astronauts to the Moon were disappearing into the past. The 1960s had been an exciting time. Fast combat aircraft, jet air travel, spacemen waving from orbit and astronauts bunny-hopping on the Moon pointed towards a fantastic future of unlimited growth with limitless natural resources. Ahead were bases on the Moon and missions to Mars. The future was bright.

Nothing, it seemed, could challenge American technology, threaten it from competition or topple it from the global leadership role in politics, finance, and weapons. It had confronted the Soviet Union, won the Space Race and was heading to defeat communism on both hemispheres. It was a great time to be alive – in the USA. Counter-intuitively, from a seemingly unassailable high ground of confidence and optimism, by the end of the decade the mood had begun to change.

Tired of conflict in Southeast Asia and responding to a catalogue of environmental warnings and protests against over-industrialisation, the American people responded positively to a seminal drive to ban toxic chemicals and clean up the planet. A wave of books drawing attention to the dangers of hubris in the goal for US supremacy vented public dissatisfaction over increasing investment in major technological programmes and the ever-expanding drive toward indiscriminate growth.

Standing out against further abuse of the Earth and its environment was Barbara Ward's book 'Only One Earth'. When it appeared in 1972, its publisher put the whole-Earth picture taken by Apollo 17 on its return from the last Moon landing on its cover, an indicator of a message for sustainable co-existence with the planet. This resonated with electors who sent their political representatives clear messages of support.

Recognising the prevalent mood in Congress and the country, NASA's Earth resources research programme quickly became the agency's flagship post-Apollo public-relations campaign. It was a vote catcher and got the attention of Congress, a feelgood factor in a world divided by political ideology and a war in Southeast Asia. It gave the future for space programmes a new vitality and there were many in State legislatures that saw that as beneficial to the electorate and the economy.

Between 1969 and 1974, NASA survived the challenging years of the Nixon administration and an increasingly cynical electorate questioning the drive towards further technological supremacy on the global stage. In this period

NASA gained approval to build the Shuttle, ostensibly to provide economical access to space, turning its focus to a different planet seemingly lost in the rush to other worlds – the Earth and its rapidly changing environment.

High level of support

By the mid-1970s, NASA's remote-sensing satellites had built a high level of support from a wide range of users and applicators. Benefitting most were land-use assessors, agriculturalists, and geophysicists across the United States and increasingly around the world. The media too became interested. With Moon landings a thing of the past, there was a completely different orientation to the space programme, and it proved popular with a public, who in large numbers had expressed their discontent with government spending on ever more ambitious deep space exploration by astronauts. In the final year of the Apollo programme, only 25 per cent of voters wanted Moon missions to continue.

The broader appeal of the Landsat programme gained the attention of the United Nations, with NASA's overall technology development role providing pointers towards a semi-operational system. NASA investment in the Landsat survey programme went from $2.3million in 1969 to almost $62million in 1978 ($320million in 2023 dollars), only a small fraction of the total global programme for Earth resource analysis from space. And it would only continue to grow.

Nevertheless, as effective as Landsat data was, the uptake was slower than the Office of Management and Budget (OMB), the former Bureau of the Budget, wanted to justify a fully operational system which would require substantial government funding. In October 1978, President Jimmy Carter tasked NASA and the Department of Commerce to find ways of stimulating greater use of space-based remote sensing and a move toward privatisation of Landsat. Just over a year later, NOAA was given temporary responsibility for managing an operational system. During this period, NASA launched additional Landsat satellites with a new instrument, the Thematic Mapper, but further development was placed on hold pending the outcome of privatisation.

In 1981, the newly installed president Ronald Reagan pushed the idea further and strongly supported the urge to remove Landsat from government ownership. Two years later, he announced the decision to transfer Landsat and weather satellite programmes to the commercial sector, but the Department of Commerce pressed ahead with privatising just the Landsat system.

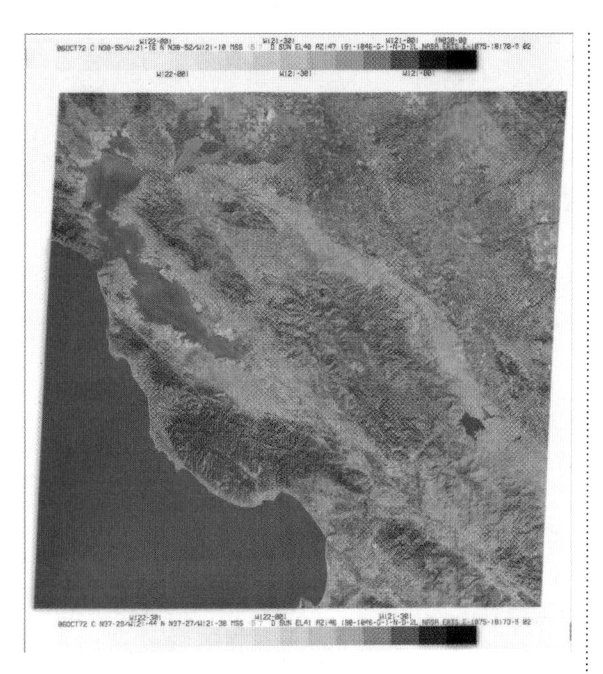

LEFT • From 1972, the first Landsat 1 images provided data for deciding on appropriate instruments for later satellites, expanding their potential. (NASA)

In 1984, a joint venture between Hughes and RCA set up the Earth Observation Satellite Company (EOSAT) for data dissemination with Congressional legislation to continue limited government support for further development. It would compromise decisions over future Landsat satellites, particularly with the next phase of development – Landsats 4 and 5.

Landsat 4 was altogether different to the first three experimental types, a much larger satellite built by GE Astro Space as a development on preceding satellites in the series but with an added requirement. It had to be recoverable, repairable in space and capable of filling different mission needs. At the core of its post-Apollo

BELOW • Located in Greenbelt, Maryland, NASA's Goddard Space Flight Center became the focal point for Earth and environmental studies with Landsat and, from the 1970s, a growing flotilla of space-based platforms for intensively studying the Earth. (NASA)

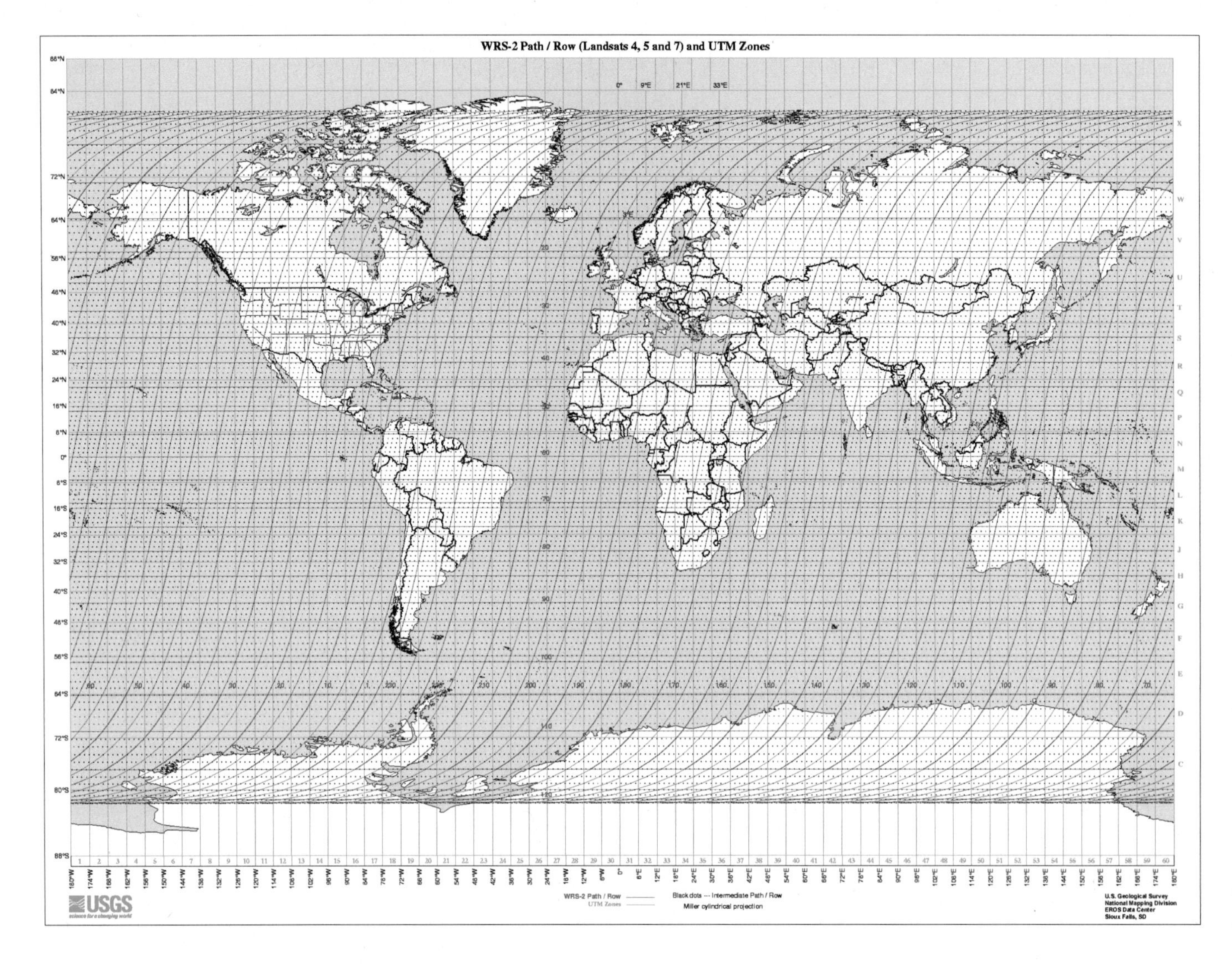

ABOVE • The Worldwide Reference System was set up to provide a global grid of geographic locations for reference against Landsat images, enabling an inventory for users around the world. (NOAA)

programme, NASA wanted to cut the costs of space operations and the Shuttle had been a part of that. Given that most satellites could be grouped into specific categories, the 'bus' part which would support the payload could be made to a common design concept. Only the payload would be bespoke to a particular mission requirement.

Universal bus

This universal 'bus' was named the Multi-mission Modular Satellite (MMS) and NASA wanted to have it available as a low-budget procurement for different satellites to take advantage of the Shuttle's ability to rendezvous with objects in space. The MMS would have grapple fixtures allowing the Shuttle's manipulator arm to capture it for refurbishment or repair. The first to use the Fairchild-built MMS platform was NASA's Solar Maximum Mission satellite, which astronauts would rendezvous with and repair in April 1984.

Landsat 4 was the second mission based around the MMS concept and was a considerable upgrade on its predecessors, with a weight of 1,941kg (4,279lb) and a single articulated solar array panel producing 1.43kW of electrical power. It had a hydrazine attitude control system and momentum wheels for stabilisation and pointing to within 0.01 degree of the desired angle. In a lower altitude of 705km (438 miles), Landsat 4 would have a higher field of view to retain the same swath of 185km (115 miles) along the ground track.

Launched by Delta 3920 from Vandenberg Air Force Base on July 16, 1982, Landsat 4 had the RBV cameras replaced with a much more effective Thematic Mapper which was introduced on this satellite for evaluation. It had seven bands, three in the visible wavelengths and four in the infra-red with a resolution of up to 30m (98ft). Readers will recall the limit placed on resolution by the intelligence community to prevent the use of Landsat images by unfriendly states for military purposes. That had been relaxed after international competitors produced satellites with higher resolution, threatening to upstage Landsat as recoded in the next chapter.

RIGHT • The first Landsat 2 image of Alberta, Canada, which provided the basis for a national environmental development programme that encouraged government planning into relieving pressures on rural communities. (NASA)

When Landsat 4 was launched, NASA had a plan to retrieve it on the second of a planned series of Shuttle missions from Vandenberg and return it to Earth for refurbishment, a quality example of how the agency wanted future uses for the Shuttle. It was for this reason that the satellite was placed in a lower orbit, more accessible for the Shuttle. NASA intended flying some polar-orbital Shuttle missions from the California launch site, including the capture, and replenishing of spy satellites. The Landsat 4 recovery flight was planned for April 1986 but delays to the Shuttle programme extended that date, and it was finally abandoned. All Shuttle flights from Vandenberg were cancelled after the Challenger disaster of January 28, 1986.

Building Landsat 4 for in-orbit repair or return to Earth for refurbishment was only one of the changes affecting the programme. In developing the manned Space Shuttle, NASA planned to replace many ground stations with a system where communications with the crew could be maintained continuously. This required a relay platform in geostationary orbit 35,880km (22,300 miles) above the Earth, receiving signals from the Shuttle at low orbit and re-transmitting them to a ground station continuously within radio sight of the satellite.

In such a system, the Shuttle and other satellites would first send their signals far out to the relay platform, which would amplify the signal and retransmit it to a ground receiving station at White Sands, New Mexico. From there, the information would be distributed to wherever it was needed. This would eliminate most US ground stations around the world and would be known as the Tracking and Data Relay Satellite System (TDRSS). It would initially consist of two satellites separated by 130° in longitude. Later, TDRSS satellites would be located at three positions around 120° apart around the globe for total and unbroken communications with all satellites and spacecraft.

Late in space

The first TDRSS was late in being sent into space and not until 15 months after the launch of Landsat 4 did it reach orbit. Designed to send data directly to Earth via TDRSS, problems with the on-board tape recorders left the new relay link as the sole means of transmitting data from the Thermal Mapper. Landsat 4 had to wait until the relay became operational on August 12, 1983. But it worked and presaged a new era in which continuous data was possible and not reliant on the troublesome recorders. TDRSS has grown and expanded into a comprehensive data relay network serving a wide range of satellites in orbit.

There was another unique feature for Landsat 4 operations. As noted later, the development of a satellite-based navigation system had been underway for some time at different military agencies and the semi-operational use of such a system was already being trialled in the civilian sector. Highly accurate position information is key to obtaining optimum use of the data from Earth-observing satellites. Landsats 4 and 5 pioneered the use of the Global Position System (GPS) for precise location of the satellite in space with respect to the Earth's surface. At the time, only four of a planned constellation of 18 GPS satellites were in orbit and there were times when Landsat 4 had no GPS signal at all. As the system evolved, performance improved and a report quoted accuracies of 50m (164ft) even with the partial and semi-operational system.

It was a timely upgrade in capabilities. The Landsat commercialisation agreement of July 17, 1984, stipulated that the data archive be the preserve of the Commerce Department and prevented EOSAT from claiming exclusive data rights. This cleared the way for operation of Landsats

ABOVE • NASA scientist Robert Watson warned Congress in 1986 that the temperature of the Earth was rising and used Landsat data to qualify his statements. (Author's collection)

ABOVE LEFT • The last of the satellites managed exclusively by NASA, Landsat 3 was launched in March 1978 and would continue to operate for more than five years. (NASA)

BELOW • Determining the environmental consequences of Desert Storm, the military operation in 1991 to evict Iraqi forces from Kuwait was aided by data gathered by Landsat satellites. (USAF)

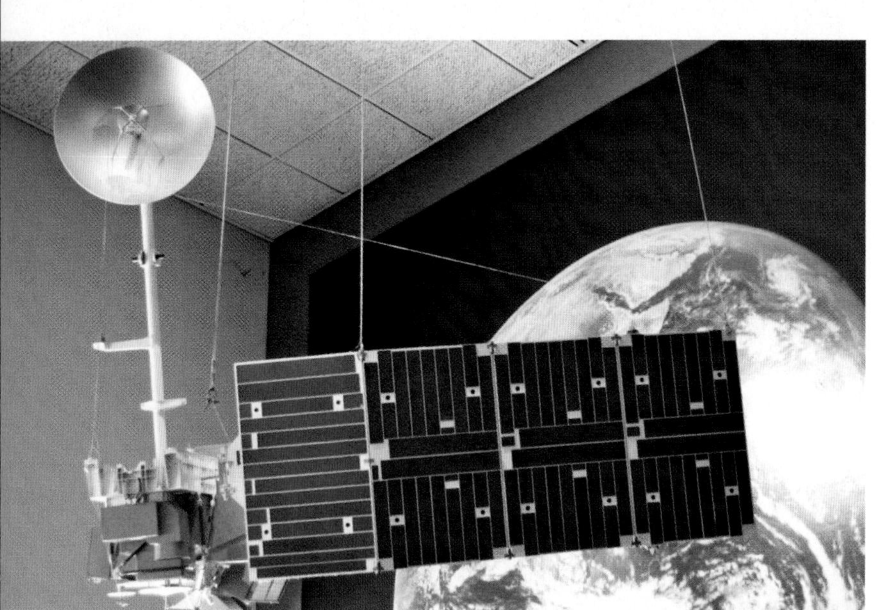

ABOVE • Built by General Electric, later absorbed into Lockheed Martin, Landsat 4 was launched in 1982 and operated for almost 19 years. (NOAA)

BELOW • Geologic features and landform folds in Iran provide detailed information on ground materials and help resource analysis for national inventories. (NOAA)

4 and 5 and to plan for Landsats 6 and 7. Highly supportive of commercialisation in several satellite programmes now operational and far beyond being mere technology demonstrators, the resolution limit for Earth observation satellites was reduced to 10m (32.8ft) which recognised competition coming from other countries.

Transformational change

An identical twin of its predecessor, Landsat 5 was launched on March 1, 1984, with an MSS and a Thermal Mapper. It would provide outstanding performance and deliver more than 2.5 million images during its 28-year life. There were technical issues and problems when the articulated solar array caused a suspension of operations until new operating procedures proved successful. The transmitter also proved troublesome and there were serious issues with the travelling-wave tube amplifiers, but the results of the instruments were outstanding.

Landsat 5 was the first satellite to document the Chernobyl nuclear incident in 1986 and played a major role in mapping areas of deforestation in Southeast Asia. One asset of its design followed that of Landsat 4 which provided additional propellant so that the orbit of the satellite could be lowered to that of the Shuttle, as originally planned and described above. Of course, that was not necessary, so the excess was used to keep it operating in its existing orbit and efficient use of the attitude control system transferred the advantage to longer life. Landsat 5 sent its last image on January 6, 2013, with the final transmission six months later, 29 years three months and four days after launch.

As reviewed in the following chapter, attention to the Earth's changing environment began to take effect during the early 1980s. It started with the prolific abundance of satellite imagery which began to expose some uncomfortable truths. That damage to the Earth's fragile ecological structures caused by indiscriminate development and unconstrained expansion of the human population was bringing demands for moderation in growth that could not be addressed by existing policies. What had been apparent, and accepted by many politicians in the 1970s, was now seen as a potential catastrophe threatening the status quo.

Having placed support for Earth and environmental research at its core and with experience working these issues with the USGS, NOAA and other interested government agencies, NASA was in the vanguard of environmental and climate research. As early as 1983, it linked up with the US National Center for Atmospheric Research to add weight to NOAA's Climate and Global Change programme. By the end of the decade NASA, the NOAA, and the National Science Foundation (NSF) had formed a tripartite assault on ignorance and misinformation from companies and organisations that sought to downplay or minimise the risks to Earth, life, and human prosperity. Many vested interests, private companies and corporations resisted the evidence from sustained monitoring of the Earth's environment which provided irrefutable scientific data of unsustainable growth and permanent environmental harm.

Skilfully, the tripartite group convinced the budgeteers in the Reagan administration to accept the evidence and to turn government agencies towards a more efficient way of handling the flow of information and acting upon it. It was in time to influence the 1989 'Presidential Initiative – The Global Climate Research Program'. During hearings in the US Senate in 1986, NASA's Robert Watson provided evidence to back up his claim that 'global warming' was inevitable and that: "It is only a matter of magnitude and timing." In that year also, the UN mandated that remote sensing was the primary tool for understanding the scale of threats and providing robust evidence-based science.

Ozone depletion

Only the previous year, British scientist Jonathan Shanklin had conducted research into the ozone layer over the Antarctic continent and discovered that in 20 years, it had been significantly depleted. The deterioration was caused by the widespread use of chlorofluorocarbons (CFCs) in the stratosphere and the US government began the process of banning those throughout industry and in domestic goods such as refrigerators and freezers. They would not be banned in Europe and the UK for a further 15 years.

In support of these pioneering measures, astronaut Sally Ride, the first American woman in space, pressed NASA to

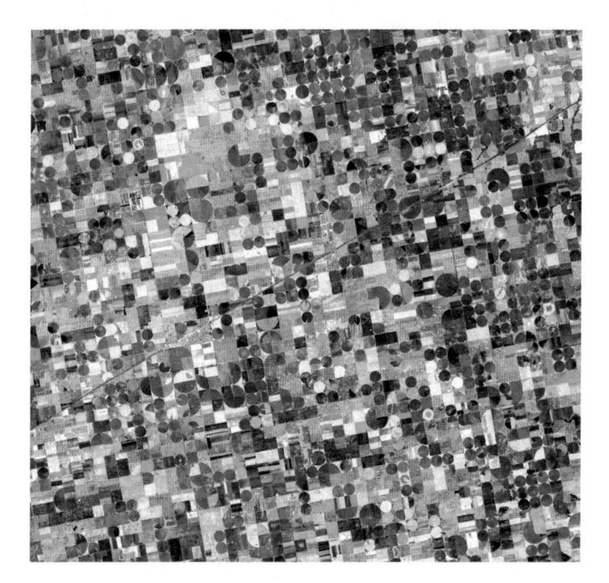

place its 'Mission to Planet Earth' programme as its highest priority. In that, the US was instrumental in forcing through the Montreal Protocol pressing nations to follow suit and ban CFCs which was enacted on January 1, 1989. Hailed as one of the most egalitarian acts of nations for the common good, it would only become law in the UK in 2000. The message had been given energy in 1988 when Congressman George Brown Jr, asserted: "Our inability to forecast the implications of human induced climate change stems from our vast ignorance of how...we are disturbing our surroundings."

Where once it had been mainstream to ridicule the doomsayers and those professing to warn of environmental collapse, it was now government policy to legislate against those triggers and to move positively to use satellite data to provide timely information. Shanklin's discovery about the ozone layer was made using such data and the sustained monitoring by satellites to measure progress in mitigation efforts persists to this day. What emerged from all this pioneering work was the US-led Global Change Research Program (GCRP), which under successive presidents became law on November 16, 1990.

Perhaps paradoxically, the Pentagon's use of Landsat data surged with Operations Desert Shield and Desert Storm (August 1990-February 1991) to evict Iraqi forces from Kuwait. The spectral range of Landsat data significantly improved the picture of land areas and surface types better than dedicated military satellites or foreign satellite imagery from Spot Image, more about which in the next chapter. Landsat data provided engineers with data for planning air bases, spectral analysis showing different soil and sand conditions and the preparation of maps for roads and convoy routes.

During these operations, EOSAT prioritised data delivery to the military and helped the Pentagon appreciate the diversity of products available from the civilian sector. Thus began a slow migration from specific military satellite programmes to commercial data procurement, opening a wide and expanding private market. During Operation Desert Storm, it was the only information the Pentagon could share with coalition forces, comprising several military units from Middle East states. The continued presence of coalition forces in Iraq extended the use of this data for many years.

No easy job

Management of Landsat had never been easy and the balance between the public and private sectors was uncertain, lawmakers never happy with the requirement for public funds to support a commercial marketing exercise. It began to unravel in the early 1990s when congressional hearings tackled enduring problems.

A significant issue was the amount of data purchased and the cost of developing, building, and launching the satellites. That issue came to a head in 1992 as final preparations were underway for the flight of Landsat 6. Congress accepted that the bias towards the private sector had failed, and it agreed that Landsat should pass back into government ownership, leaving EOSAT to market the data only.

But even as the legislative framework for a new kind of ownership and shared data handling was sorted out, another unexpected disruptor arrived. The Internet. Throughout the early 1990s the World Wide Web grew exponentially, and users of Landsat data strapped together new and exciting opportunities to distribute information obtained from images delivered by Landsats 4 and 5 and through the expanding databank of historic images. New ways of exploiting the multi-spectral data emerged in ways never previously envisioned and for purposes now shaped by the preoccupation with extreme weather and climate change.

In other ways too, the 1990s reshaped the entire spectrum of funding and support for NASA programmes. Where at the beginning of the decade manned space flight accounted for virtually half of all NASA activity, by 1999 that had fallen to little more than a third, while funding for science and technology had increased from less than a third to almost a half. The total workforce at NASA was reduced by a third and the contactor workforce by a half.

To those closely involved and affected by NASA programmes, it was a time of unprecedented change, a redirection of what the space agency aimed to achieve through policy and programme plans. Instead of pursuing exploration, it was now an agency for general applications, for better management of the Earth and its finite resources, for a stronger connection between high science and everyday benefits.

The Shuttle opened unique possibilities, for use as a laboratory in space, for launching Earth satellites and for providing a seemingly routine access to space. During the 1990s, a new age of environmental awareness influenced major space policy decisions and took politicians to new realms of understanding, frequently outside their own comfort zones. It stimulated a wide range of global space ambitions as new possibilities in many countries focused on how they could use space for doing unimaginable things – such as increasing food production and crop yield through better management of agriculture, by better balance between urban and rural planning and through an inventory of natural resources.

LEFT • *Landsat was effective in aiding agricultural programmes and provided farmers with valuable data on the health of their crops, as shown here. (NOAA)*

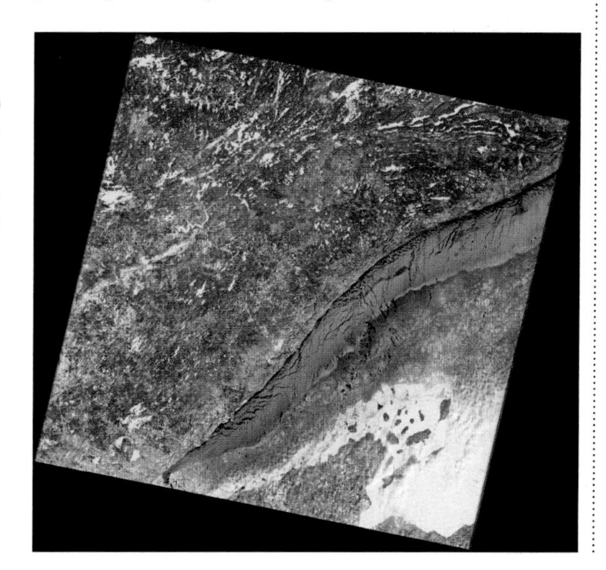

LEFT • *This view of ice fields on Lake Superior provided by Landsat 5 helped planners determine the resources required to clear routes for free passage. (NOAA)*

A DEEPER LOOK AT THE EARTH

News of developments under way with commercial and civil programmes.

ABOVE • The Enhanced Thermal Mapper during development for flight on the Landsat 7 mission, creating new capabilities for interpretation. (NASA)

Throughout the transformation of the Landsat programme and its expansion from a tool for agricultural surveys to a platform for monitoring changes to the Earth and its environment, the next generation satellites were in development. The new organisational structure had satellites funded by the government once again, designed and developed by a contractor, built, and launched by NASA, managed in its operations by NOAA and with data distribution through EOSAT.

Landsat 6 was sent up from Vandenberg on October 5, 1993, by a Titan 23G rocket, a retired Titan II ballistic missile adapted for space launches. Much was expected of this satellite, which carried an eight-band Enhanced

LEFT • Vandenberg Air Force Base, California, from where all polar-orbiting satellites are launched, as viewed by Landsat. (NOAA)

Thermal Mapper including a panchromatic band with 15m (49ft) resolution and a thermal band. Manufactured by Martin Marietta, it had a launch weight of 2,200kg (4,800lb) and included 122kg (270lb) of hydrazine propellant. Shortly after separating from the second stage of the rocket a manifold in the hydrazine line ruptured and the satellite was unable to maintain attitude control and rapidly became unusable.

With Landsat 7 five years away from launch, only the ageing Landsat 5 could maintain supply of data for EOSAT, that organisation now a joint venture between Lockheed Martin and Hughes. By 1996, EOSAT had been sold to Space Imaging, a company that was already marketing satellite data from India, Russia, Canada, Japan, European countries and from a range of airborne image providers. With Landsat no longer the exclusive preserve of remote-sensing data from space, the world had responded to the US-led drive for environmental concerns with global management of climate change and the threats that those posed.

It was also a time when the defence establishment and the intelligence community began a transition away from big-ticket programmes for massive spy satellites and expensive optical platforms in orbit. Time to respond to commercial and civil satellite programmes which, by the 1990s were selling on the open market products that equalled, and sometimes exceeded, the performance capabilities and the product range of multi-spectral data from government satellites. Unleashed by the potential on Landsat, foreign competitors were lining up for a piece of the market.

An international armada

During the late 1970s, the French national space agency (CNES) began work on remote sensing satellites. The lead role taken by Landsat attracted competition for lucrative markets beginning to open up as a direct result of the

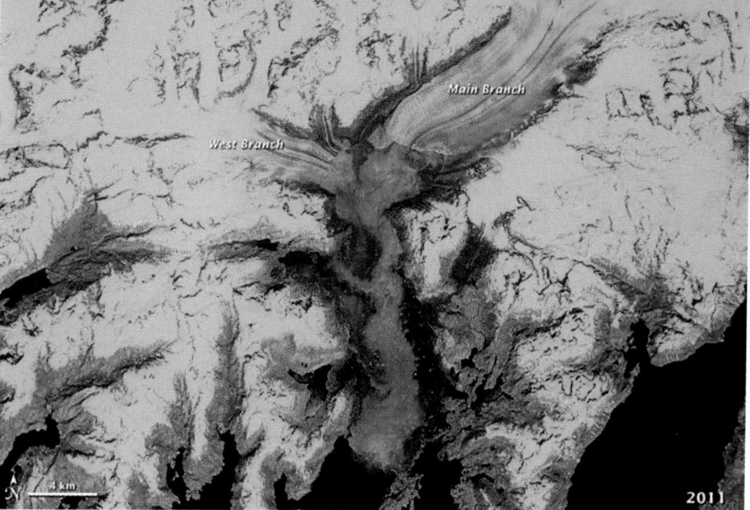

BELOW • Comparative views by remote-sensing satellites several years apart reveal subtle changes to the Columbia Glacier, Alaska, in 1986 (top) and 2011. (NOAA)

ABOVE • *Operated by the US Geological Survey, Landsat 7 in preparation for launch in April 1999, as technicians work on spacecraft systems and payload instruments. (USGS)*

remote-sensing data. France wanted to take the lead for Europe and sought a national role in using the expertise of its satellite manufacturers and the European Space Agency's Ariane rockets to deliver an optional alternative.

The consortium of US government agencies and private partnerships that underpinned Landsat were constrained by technical specifications compiled by the operators and these were not aligned with commercial demands. France saw a niche, providing higher resolution images than those allowed under US constraints. With Ariane proving a reliable launch system and armed with solid market analysis as to global requirements, 'Systéme Probatoire d'Observation de la Terre' (SPOT) was formed by the French government in February 1978 in Toulouse, France, during 1982 and a year later an office opened in the USA.

Development of the first satellites in the SPOT series was undertaken by a consortium of France's aerospace companies with government management of specifications. The initial plan was to launch three identical satellites which would carry similar instruments. Each would have a multi-spectral instrument with three spectral bands covering green, red and near-infra-red and composites producing images with a resolution of 20m (65ft). The second instrument would provide images in a single black and white panchromatic band with a resolution of 10m (33ft). This was far greater than available by any other system at the time and attracted high interest. Both instruments were known as High Resolution Visible (HRV) instruments.

The operating orbit was similar to that of the Landsats, optimised by the rotation of the Earth and the physics of space. But the lower altitude offered a higher ground resolution, on which basis the products were marketed well in advance of launch and in support of a global sales effort. Breaking through the resolution barrier set by the US government proved the biggest pull for users. Stereo imaging was the second attraction, and, in both areas, the French programme found itself complementary to Landsat, which was still ostensibly a research and development programme.

To achieve the desired 10m resolution over a ground swath 60km (37 miles) wide could only be achieved through an array of 3,000 detectors for each spectral band, sampling every three milliseconds. The 10m resolution called for 6,000 detectors per line sampled every 1.5 milliseconds, both requirements met by charged-couple devices (CCDs). In setting the imaging sequence, the panchromatic channel provided a field of view constantly 15km (9.3 miles) ahead of the colour scan swath.

In marketing the capability, SPOT emphasised the applications for spectral band applications, identifying the green band tuned to chlorophyll and capable of penetrating surface water to a depth of up to 20m (66ft) in clear conditions. The red band was similar to the Thematic Mapper design for Landsat 6 and would help identify crops, bar oil and rocky surfaces and for water penetration to 2m (6.5ft). The near-infra-red band would show clear identification of vegetation while the red and near-infrared bands combined would help identify the quantity and volume of biomass.

The first SPOT satellite would not launch before 1986 but there was already a latecomer to the marketplace in the unlikely form of the Russian cartographic institute, Soyuz Karta. Under an agreement with the Kremlin, this arm of the Soviet government was testing its ability to enter Western markets and sell its own Earth images from retired first-generation spy satellites which had been declassified in favour of later technology. Soyuz Karta was responsible for all the maps produced in the Soviet Union and that included those produced by satellite images.

What made this very different from the civilian Landsat and SPOT satellites was the nature of the product. They were all large, wet-film photographs returned to Earth by a derivative of Russia's Vostok spacecraft responsible for carrying six Soviet cosmonauts into orbit between 1961 and 1963. Known as Zenit, it had been used for carrying into orbit large cameras fixed inside the recoverable capsule to photograph the surface of the Earth through an optical window. The process involved returning the entire spacecraft where the film would be retrieved and subsequently processed.

Different path

The Soviet spy satellite programme had been very different to that in the United States. The first and second Russian satellites had been hurriedly prepared to upstage America with flights in 1957, while Sputnik 3, approved in 1956, had

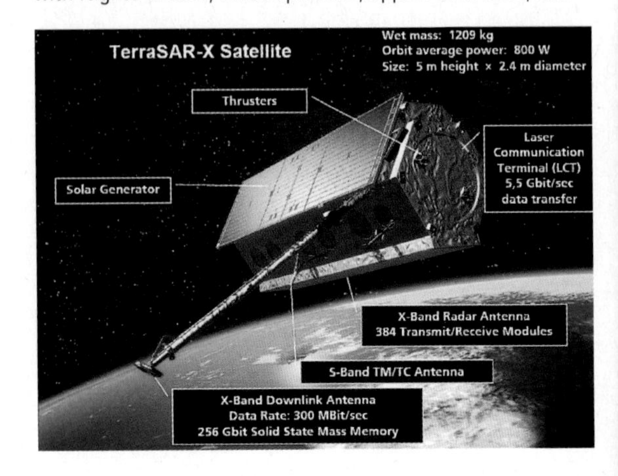

RIGHT • *Using synthetic aperture radar, two satellites are capable of providing a bi-static image in 3D of surface features, a capability provided by the TerraSAR-X satellite which was paired with TanDEM-X for this purpose. (DLR)*

a protracted development time and would not fly until 1958. Known as Object D, it was in parallel with a manned satellite programme designated Object OD-1. Russia had been experimenting with biological samples carried on sounding-rockets and wanted to extend this into the flight of a man in orbit.

When it ran into technical difficulties, work began on its successor but to avoid possible cancellation, design engineer Sergei Korolev proposed Object OD-2, approved in May 1959, which would be capable of carrying a cosmonaut in the K-2 variant or a spy camera in K-3.

The same basic design of spacecraft would be used for both applications and that gave Korolev assurance that, with the priority at the time going on military projects, he could still count on support for his real ambition to put a man in space. There had been significant hostility to that from the military, criticising Korolev's 'frivolous' use of national resources. Newly won over to the propaganda value of space activity, Khrushchev had backed him all the way.

As designed, the OD-2 consisted of a near-spherical re-entry capsule with a diameter of 2.3m (7.5ft) weighing 2,400kg (5,300lb) attached to a conical service module, the combination weighing a total 4,700kg (10,363lb). This capsule had a pressurised atmosphere of oxygen and nitrogen and the service module provided electrical power from batteries, thermal control for the re-entry module and attitude thrusters for pointing accuracy, together with a retrorocket for coming back to Earth.

The advantage with this concept was that standard camera equipment developed for high altitude aerial photography could be carried in the pressurised, temperature-controlled interior without any special requirement for the sensitive equipment surviving in a vacuum, unlike the US Corona satellites, where the cameras and film were in an unpressurised capsule. It was a simple solution with the broader programme serving two objectives, the political propaganda of a man in space using the K-2 (Vostok) and a spy satellite for the K-3.

The spy satellite variant was named Zenit-2 and could provide high resolution images, routinely down to 5m (16ft) and sometimes to 1m (3.3ft). When the main re-entry capsule returned to Earth, it brought back the total exposed film load and the camera too, which was a great advantage with its simplicity of operation and cost-effectiveness. The orbit planned for most Zenit 2 flights was about 200km (125 miles) by 250-350km (155-217 miles) at an inclination of 65° to the equator which was product of the location of the Baikonur launch complex at 63.5° north latitude.

The race for data

Identified by the Russians as Korabl-Sputnik 1, the first K-2 prototype Vostok spacecraft was launched on May 15, 1960 and recovered on September 5. It failed to achieve the required attitude orientation and when its de-orbit motor fired, it went into a higher orbit, from which it decayed naturally back through the atmosphere. The second Vostok was launched on July 28 with the dogs Chayka and Lisichka but that blew up shortly after launch. It was followed on August 19 by the third Vostok carrying the dogs Strelka and Belka, which were recovered safely from orbit after 14 hours. Strelka had puppies a year later, one of which was sent as a goodwill gift to Jacqueline Kennedy shortly after her husband became president.

Four more were tested, of which only two were a success, before the launch of Vostok 1 carrying Yuri Gagarin on April 12, 1961. Five more flights with cosmonauts followed, the last on June 16, 1963, carrying the first woman into space when Valentina Tereshkova remained in orbit for almost three days. Russia's space plans were evolving, and changes were frequent. Originally planning for a further seven Vostok missions, the basic spacecraft was developed into the Voskhod variant using essentially the same hardware with provision for up to three people.

Only two Voskhod flights were made, the first on October 12, 1964, when three cosmonauts spent a day in

ABOVE • *Landsat 9 in the final stages of checkout prior to launch on September 27, 2021. (Northrop Grumman)*

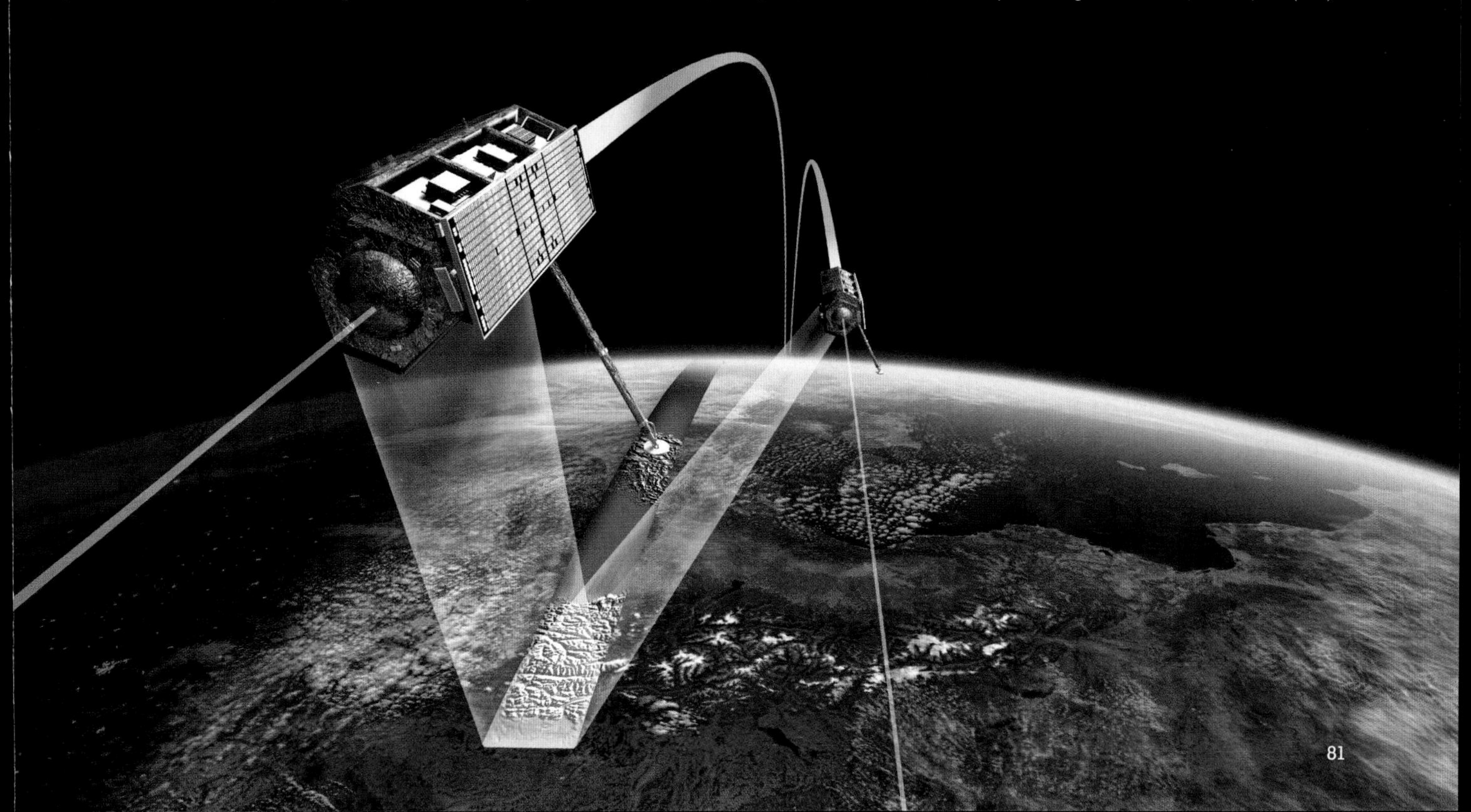

BELOW • *The paired satellites TanDEM-X and TerraSAR-X are flown just 350m (1,148ft) apart to obtain bi-static images of the surface. (ESA)*

space and the last on March 18, 1965, with two cosmonauts, including Alexei Leonov who made the first spacewalk. The Vostok/Voskhod flights were over, and attention turned to a Moon landing programme involving Korolev's giant N-1 rocket and the Soyuz spacecraft, the latter carrying Vladimir Komarov to his death when the first manned Soyuz crashed on April 24, 1967.

All this while, military engineers and scientists had been developing the Zenit 2 spy satellite variant of the Vostok spacecraft. The first flight attempt on December 11, 1961, failed when the upper stage of the launch vehicle malfunctioned, and operators triggered the self-destruct package to prevent it re-entering and falling into unfriendly hands. The first successful flight began on April 26, 1962, when the publicly designated Kosmos 4 was launched. Its primary function was to monitor radiation after US nuclear weapons testing in the Pacific Ocean, but it failed to produce optimum photographs due to an attitude orientation problem.

The next launch attempt caused widespread damage to the launch pad on June 1, 1962, when one of the four strap-on boosters fell away and exploded, with the rest of the rocket falling a short distance away. Repairs to the pad delayed the joint flights of the third and fourth manned Vostok missions to mid-August but the next Zenit 2 flight on July 28, 1962, was a success. The last of 81 Zenit 2 flights was launched as Kosmos 344 on May 12, 1970.

The Zenit 2 series carried the Ftor-2 package, which incorporated five cameras inside the pressurised cabin. Four had a focal length of 1000mm with a single low-resolution camera of 200mm focal length for a large-area view to place the high-resolution photographs in context. Each camera had the potential for a maximum 1,500 frames, presenting an image covering an area of 60km x 60km (37.3 miles x 37.3 miles). At operating altitude, this provided images down to a resolution of around 10m (32.8ft), with some flights achieving a resolution down to 2m (6.4ft).

In addition, the Zenit 2 flights carried Kust-12M electronic intelligence (ELINT) equipment to gather signals information over NATO countries, a requirement considered equal in importance to the photographic frames. Special antennas were attached for gathering this information, which was recorded on board and analysed after recovery. In addition, the information thus obtained proved useful in detecting coded radio and signals intelligence. Over a period of eight years, 85 per cent of Zenit 2 flights were completely or at least partially successful.

Modifications to the spacecraft and improvements to the lift capacity of the launch vehicle raised the weight of these satellites to 6,300kg (13,890lb). With greater operating flexibility, a new generation of Zenit 4 satellites appeared with the launch of Kosmos 22 on November 16, 1963, the last of 76 launches being Kosmos 355 on August 7, 1970, recovered eight days later. Flown concurrent with the Zenit 2 class, they were optimised for high resolution imagery and carried one camera with a 3,000mm telephoto lens producing images with a resolution of 1m (3.2ft) and a single 200mm camera to place the close-look pictures in context.

Improved capability

From the Zenit 2 emerged the Zenit 2M, which was the same as the basic spy satellite but with the added weight provided by the improved launch capability for the Zenit 4 and with solar panels to extend the orbital life of the vehicle. The first was launched on March 21, 1968, as Kosmos 208, with the last of 101 flights sent into orbit on August 17, 1979. Most lasted almost two weeks before returning to Earth. Many carried a special Nauka module with science instruments capable of surviving in the vacuum of space, a cylindrical compartment carried on top of the re-entry module housing different sensors and detectors.

Further parallel developments introduced the Zenit 4MK as Kosmos 317, launched on December 23, 1969, with a capacity for changing its orbital path, dipping down as low as 180km (112 miles) above the surface of the Earth to enhance a new and credible high-resolution capability. Last of 80 flights with this variant was flown on June 22, 1977. A dedicated evolution appeared on December 29, 1971, with the launch of Cosmos 470, the first of the Zenit 4MT series. These were adapted specifically to conduct mapping operations of Russia, its client states and of other countries, for which a total of 23 were launched, the last on August 3, 1982.

In a further development of the Zenit 4 series, Russia introduced the 4MKM, which began the transfer to a broader spectrum of imagery carrying special film for remote sensing. It was a significant shift to use satellites for broader advantage to the economy.

The first flight of a Zenit 4MKM began as Kosmos 927 launched on July 12, 1977, from the Plesetsk launch site and with an orbital inclination of 89.9°. Several orbital manoeuvres took place over the next few days. Six days into flight, it had an orbit of 337km x 149km (209 miles x 92.5 miles), close to the minimum height to remain in space and only just above the altitude where atmospheric molecules could slow it and pull it in. The last of 39 flights with the 4MKM was launched on October 10, 1980, with recovery 13 days later.

Experience with an increasing array of alternative Zenit satellites resulted in a shift towards multi-spectral

BELOW • *Unfolded from its packaged form for launch, NASA's Landsat 9, operated by the US Geological Survey, is the latest in a series of remote-sensing satellites first launched in 1972. (USGS)*

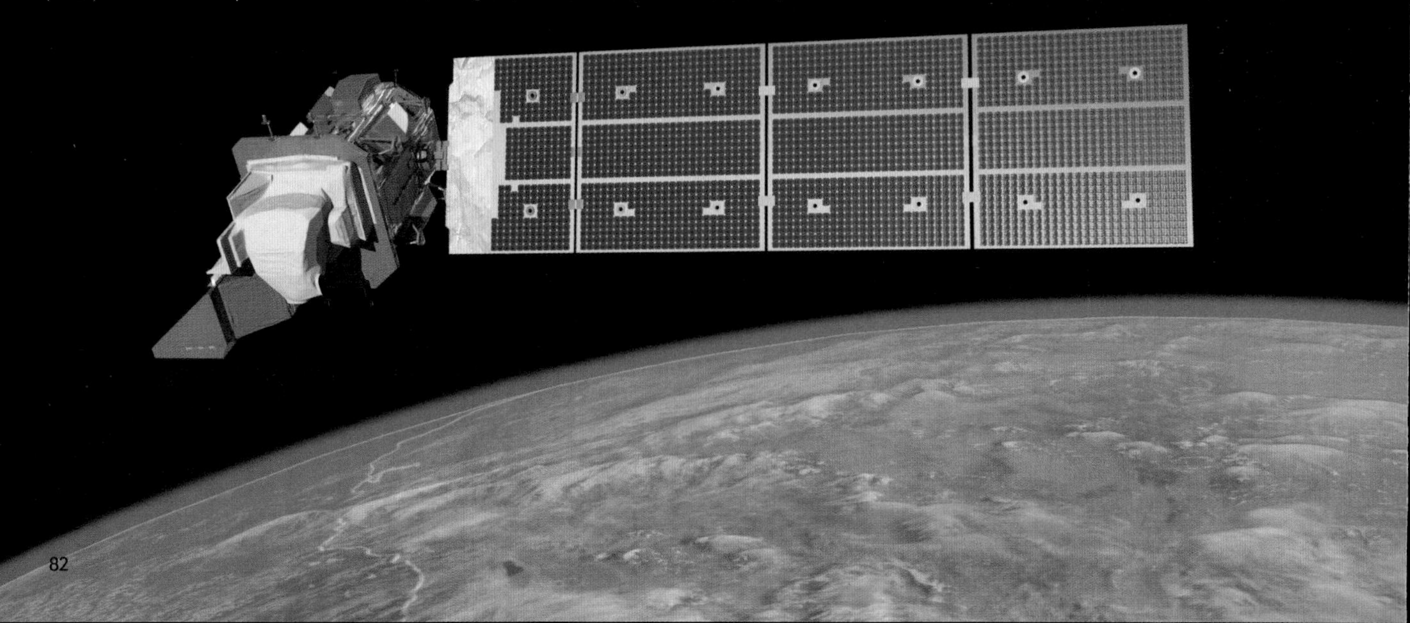

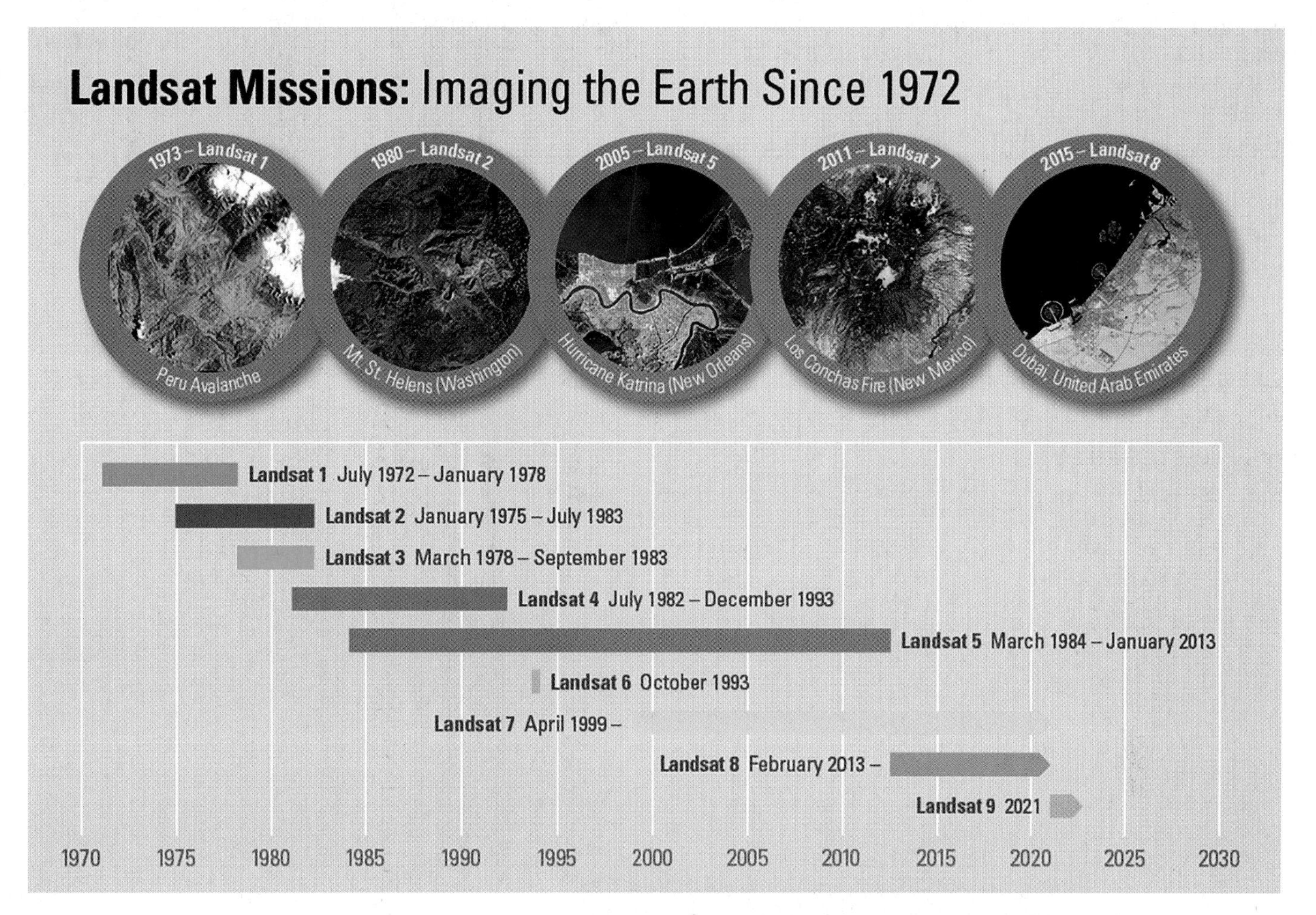

ABOVE • The timeline for Landsat spacecraft as the system and its capabilities evolved. (NOAA)

photography for Earth science, general crop yield measurement and a resource inventory. Here, as with more recent satellites in the Zenit constellation, information was applied to what would be considered the civilian sector in the West, as well as for military and intelligence gathering purposes.

Another way

Never as frequently launched as the Zenit satellites, an alternative concept was developed in which a high-resolution camera system would be flown in a structural assembly based initially on the Soyuz spacecraft then under development. But the satellite which would emerge as the Yantar-2K was produced by the Kuibyshev TsSKB, the design bureau which had taken over all Zenit reconnaissance satellite development in 1963. It was quite separate to any Soyuz derivative.

Design work began in 1964 but was frequently delayed by extensive development work on advanced manned spacecraft which never appeared and on variants of Yantar-2K which were abandoned. Multiple concepts were conceptualised and designed, only to be abandoned in a flurry of types similar to the prolific array of German combat aircraft types proposed in World War Two. And with the same result of excessive diversification and too little focus on workable types.

Zenit missions only returned film at the end of an orbital mission, usually a maximum 30 days after launch, because the cameras and film cassettes were inside the large re-entry module. Yantar-2K was designed to overcome this by carrying two small recoverable capsules for returning film separately and before the end of the 30-day mission. The structure consisted of an equipment module, a descent capsule and the two sub-capsules. The satellite consisted

LEFT • A model of the Soyuz spacecraft, designed to carry Russia's cosmonauts into space on display at the 'Russia in Space' exhibition in Frankfurt, Germany, in 2002. (de:Benutzer:HPH)

LEFT • A redundant Zenit re-entry module developed as a spy satellite from the Vostok concept shown without its thermal protection and seen in 2007 at the Russian Medical Military Academy in St Petersburg. (Maryanna Nesina)

ABOVE • *The Vostok 3KA was developed to carry cosmonauts but it served as a template for derivative variants providing Russia with its first reconnaissance satellite. (Andrew LePage/ Drew Ex Machina)*

of a truncated cone with a total length of 6.3m (20.6ft), a maximum diameter of 2.7m (8.8ft) and a weight of up to 6,600kg (14,553lb), at the base of which were two solar array wings each with a length of 6m (19.6ft).

The equipment module contained the propulsion section for changing the orbit and for maintaining attitude control, unlike the Zenit series which had no means of changing their flight paths before re-entry. The Zhemchug-4 camera equipment had a unique system for getting exposed film across to the two small re-entry capsules, each of which carried a small rocket motor for retrofire after separation and spin-stabilised for maintaining the correct orientation during descent.

The main landing capsule containing the camera equipment and operating computer was shaped somewhat like a NASA Gemini capsule and had a main retrorocket, together with attitude control thrusters, parachute systems for final deceleration and thermal protection. The operational concept for Yantar-2K was similar to the US Gambit system, whereby two recoverable capsules were deployed. Unlike its American contemporary, it also had the ability to return the camera equipment in the main recovery section so that the expensive equipment could be reused.

BELOW • *The 3KA spacecraft being prepared for launch with its thermal protection over the spherical pressurised cabin. (Andrew LePage/ Drew Ex Machina)*

Despite efforts, compromised by a diversification into multiple projects, the first flight would not take place before May 23, 1974. The upper stages of the Soyuz launch vehicle failed to separate, and it was destroyed. The second flight with the Yantar-2K on December 13 was a success but the third launch on September 5, 1975, experienced a failure on the second day when technical problems triggered the self-destruct system, a precaution against it coming back down for examination by unfriendly eyes. The first flight with the two recoverable mini capsules was launched on February 20, 1976, but they failed to get down intact. Another flight July 22, 1976, suffered battery failure and not until the sixth flight on April 26, 1977, was the type a complete success.

The Central Committee approved the programme for operations with the codename Feniks. Orbits varied between flights, but the majority followed an average 345km (214 miles) by 175km (109 miles) at orbit inclinations of 67.1-67.2°. On most flights, the two mini-capsules were separated on the 10th and 18th days of each 30-day mission. The overlap between Zenit and Yantar-2K launches satisfied specific requirements of the mission and depended upon the requirements of the military. The last of 30 flights occurred on June 28, 1983. Only two had failed.

Development of the Yantar-4K1 derivative version began in May 1977 to provide a longer duration of 45 days and to incorporate some technical improvements. The first flight took off on April 27, 1979 and lasted 30 days as planned and the second operated for a full duration from April 29 to June 12, 1980. The third flight operated successfully from October 30 to December 12, 1980, at which point the Yantar-4K1 was declared operational under the code name Oktan, in which a further nine flights were conducted, from June 8, 1982 to January 13, 1984.

Characteristic approach

The duality of role and function between civil and military application from a common set of equipment and software

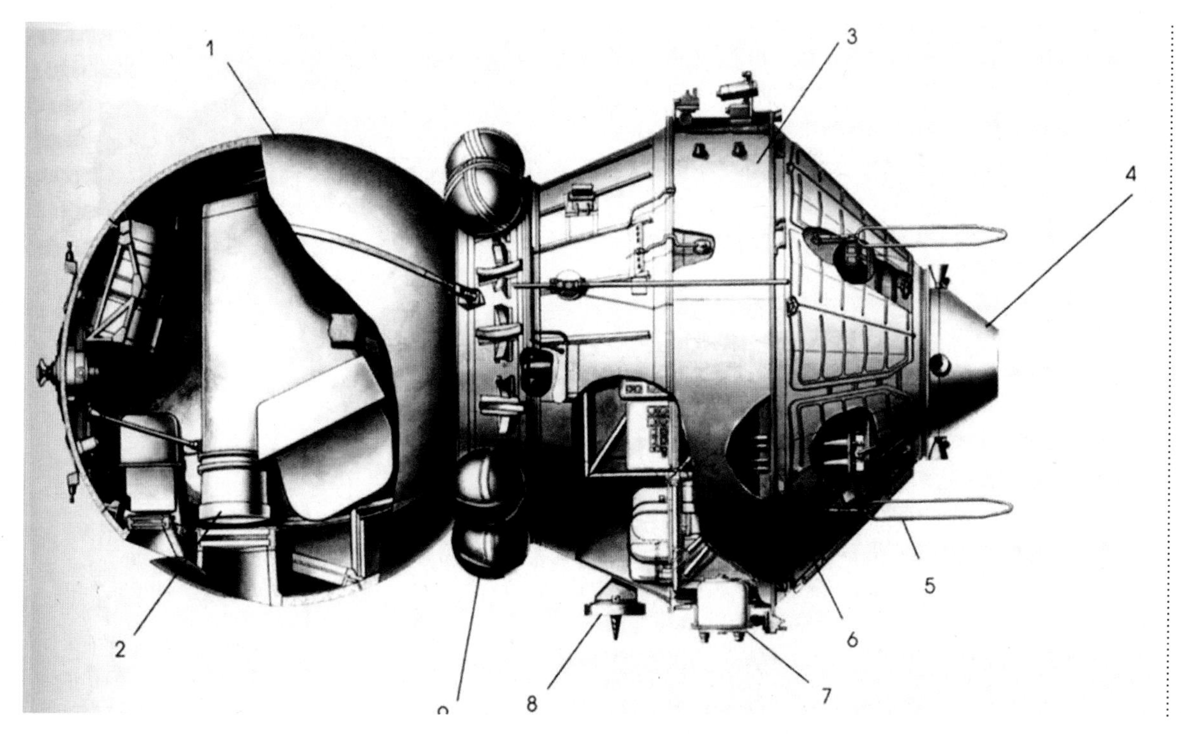

LEFT • *Each of the 76 Zenit 4 spy satellites flown carried 3000mm and 200mm cameras for high-resolution photography, the return capsule bringing back the film and the optical equipment. (Andrew LePage/ Drew Ex Machina)*

BELOW • *Development of the multi-occupancy Voskhod derivative of Vostok prompted a new and improved series of photo-reconnaissance satellites with a more powerful launch vehicle used for the Zenit 4 series. (Andrew LePage/ Drew Ex Machina)*

is characteristic of the Russian approach, even today. The enduring consistency with which they repeatedly used the same rockets, satellites and spacecraft persisted throughout the history of their space programme.

By the mid-1970s, the Russians had responded to the US Landsat series with particular attention to information they could obtain about other countries. In discussions during the mid-1980s, Soviet specialists in remote sensing confirmed that this was the time when the transition to general land management from space began to take hold in the Soviet Union.

It was marked by the first use of the Zenit 4MKT as Kosmos 771 launched on September 25, 1975. The T derivative was a specialised variant of its predecessor, the MKM, but, as a line of dedicated remote-sensing satellites, it can be considered as the founding derivative for the transition. The last of 27 such derivatives was launched on September 6, 1985, by which time the Novosti and Tass news agencies were hailing them as the new generation of Earth-observing satellites dedicated to environmental studies, to better inventories of natural resources and for 'international co-operation'.

It was in this period that the Russians decided to compete with Landsat and SPOT in marketing products from the earlier Zenit series to foreign users. The Russians had been doing brisk business with companies in the West and the country was going through a transformation in which it wanted to invest more liberally in dealings over products and an exchange of civilian technologies. There was a profound sense of change in the air and Western companies responded accordingly, with movements between vested interests in Washington DC and London and those in Moscow.

Co-operation had melted some of the ice of the Cold War. Arms control agreements had been signed, however flawed each side believed them to be. And there had been the joint Apollo-Soyuz docking mission of 1975, where Russian and American spacecraft had docked together in space, exchanged greetings between astronauts and cosmonauts and conducted a few experiments together. Later, there were even suggestions for a docking between NASA's Space Shuttle and the Russian Soyuz space station.

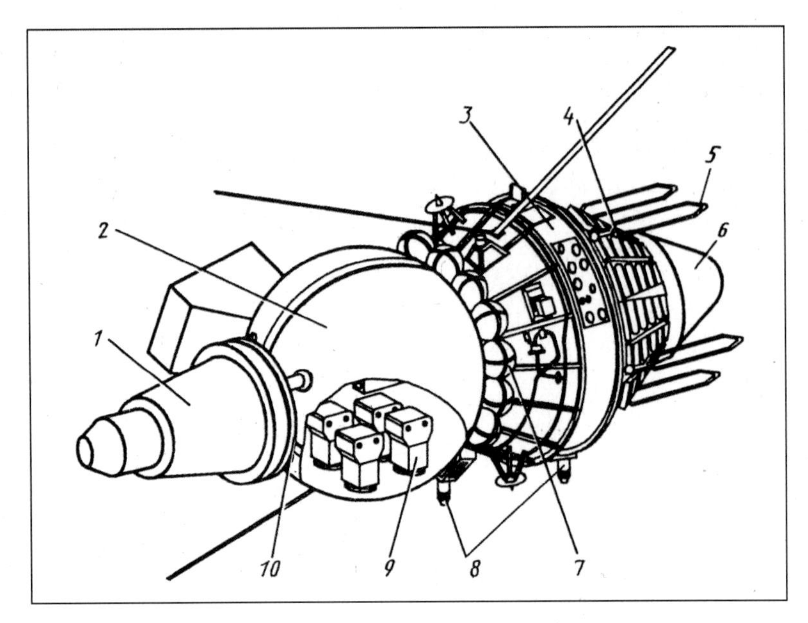

ABOVE • Under the Fram designation, Russian remote-sensing satellites were developed in the 1970s, derivatives of the Vostok/ Zenit family but precursors to the Resurs satellites. (Andrew LePage/ Drew Ex Machina)

Western companies were interested in doing business in Russia and the ability to exchange goods was increasingly possible.

Through Soyuz Karta, Russia announced interest in talking to Western companies about marketing Earth images from space. Without identifying the products as large format photographic plates from their early spy satellites, it was clear to all that this was what was on offer. Interest in gaining access to these products was beyond the obvious benefit of acquiring pictures taken from space with a resolution of 5m (16.4ft) or less. The catalogue was prepared, and Soyuz Karta produced a list of available products, the spectral bands in which the images had been taken and the prices.

Spreading the word

Companies in the United States sought to close deals with Soyuz Karta and were successful in that. But a British company with an office in the United States, London-based Space Consultants International (SCI) made several trips to Moscow and worked through various Russian government agencies to receive the distribution agreement for Soyuz Karta's Zenit products in the UK. Access to several laboratories and machine building premises in Moscow related to this work was freely given.

These were wet-film photographs of selected geographic areas on the planet and did not include territories of the USSR. The market was used now to working with digital images and to convert these positive images needed several experts, many of whom had worked on aerial photography during the Second World War, to come out of retirement and interpret or convert them to a digital format. Soyuz Karta helped in this exercise to some extent, but the magnitude of pixels required to convert a highly detailed photograph to a digital image exceeded the storage capacity of a lot of the computers operated by data users at the time.

Although some products were sold, the international market had geared up for direct reception of Landsat or SPOT images and the Russians were too late, the only marketing advantage with their film being the exceptionally high resolution. But, while fascinating to see, that resolution was too high for most users and the lack of a full, global multi-spectral survey turned customers away. That sort of product was available from other suppliers in the US and France.

Several companies approached Russian agencies with business opportunities for engaging with the West, not least the highly successful UK Commercial Space Technologies (CST), formed by Gerry Webb and which still exists. CST focused on booking satellites for launch on Russian rockets and pioneered a route to low-cost access to space for Western users. But the core of all this activity was the sustained development of new types of Earth-observing satellites.

In November 1984, chairman of the USSR State Committee for Hydrometeorology and Environmental Control said that Yury Izrael had announced that an integrated satellite-based system was being set up to monitor the environment and that this would require the use of direct-transmission of Landsat-type images to ground stations.

Long before this, the Russians launched a new variant with their Zenit 6, which first flew on November 23, 1976. It had equipment for both low- and high-resolution photography and replaced all preceding types. The last of 95 was launched on June 19, 1984, which was about the time Soyuz Karta began to market its products in earnest. Some of these images did find their way into SCI's catalogue. By this time, a successor had already flown and would be the final Zenit variant.

The military wanted a capacity to 'look' sideways from a direct view fixed to the centre of the orbital ground track, providing a capability to slew the look-angle left or right. This only became possible with the Zenit 8 variant, which was also used for studies into natural resources and environmental issues. These images did find application beyond their military application, but it was primarily operated for intelligence gathering. The first Zenit 8 was launched as Kosmos 1571 on June 11, 1984, the last of 102 in the series going up on June 7, 1994.

Earth resource satellites

For some time, derivatives of the basic Zenit were defined as Earth resource satellites in a series known as Resurs, the first of those being launched on September 5, 1979, as the F1, of which 52 were launched, the last on August 24,

RIGHT • The Yantar 2K remote-sensing satellite was in some respects a parallel to the US GAMBIT series of spy satellites in that it carried two film-return capsules and could remain in space for up to a month. (Andrew LePage/ Drew Ex Machina)

1993. Two of an unidentified configuration were also launched, in 1997 and 1999. All Resurs satellites were developed from the Zenit 4MKT series with a Priroda imaging system incorporating two KFA-1000 cameras and three KATE-200 cameras.

Both camera types had 1,000mm focal length lenses, the KFA-1000 for 1,800 frames with a resolution of up to 5m (16.4ft) and a ground swath of 60km (37 miles). The Kate-200 could collect 1,200 frames, delivering a resolution of 10-30m (32.8-98.4ft) in three spectral bands. It was possible to obtain stereo pictures from the Kate-200 by overlapping successive frames at 20 per cent, 60 per cent or 80 per cent. A star indexing camera was used for precise location.

The Resurs F2 appeared in 1987 with the first launched on December 26, the last of 10 on September 26, 1995. They were distinguished by two long solar arrays attached to the forward section of the spherical re-entry module to extend orbital life to 30 days. Each carried a SA-M imaging module consisting of four MK-4 cameras for 5-8m (16.4-26.2ft) resolution across six spectral bands. These flights straddled the period during which the collapse of the Soviet regime brought many changes to Russia, now a federation of states outside the spectrum of former Warsaw Pact countries.

Developments in post-Soviet Russia's space programme saw a shift towards digital transmission of images from remote-sensing satellites, a precursor of which had been launched on July 24, 1983. Designated Kosmos 1484, it was a test vehicle for the Resurs 01 series, the first of which went up on October 3, 1985. Only four Resurs 01 and 02 satellites were launched, the last on July 10, 1998. A range of payload options were carried, including multi-spectral scanners and radiometers for general Earth science and environmental observations. These satellites operated for several years performing a function similar to the Landsat and SPOT series, but the collapse of the USSR had a profound effect on further development.

The newly formed Russian Federation never recovered but efforts to provide data on Earth and environmental issues continued in former East European and Warsaw Pact states. In 1990, Space Consultants International, rebadged as Sigma Projects Ltd, hosted an international Earth Mission 2000 conference in London, at which cartographers, Earth scientists and environmentalists from Russia, East European and Baltic States attended to discuss opportunities. It was the beginning of a concerted effort by former Soviet-led states to finally break free from Moscow and join a growing band of scientists using satellites to study the Earth's climate and reach disturbing conclusions.

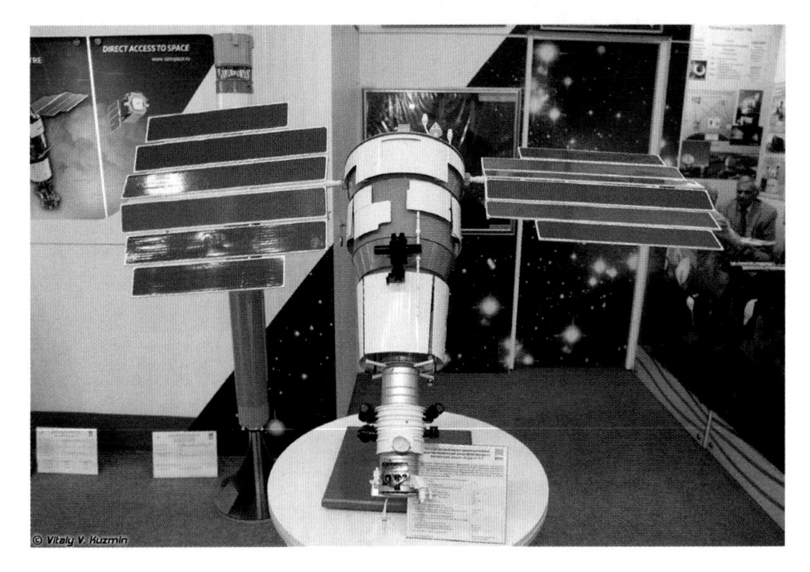

ABOVE • On view at a Moscow aerospace exhibition, Russia's 6,570kg (14,480lb) Resurs-P spacecraft series was launched from 2013 with current plans for further derivatives on hold. (Vitaly V Kuzman)

BELOW • The Foton-M derivative of the Vostok/Zenit family was used primarily for space science and biology experiments. (Andrew LePage/ Drew Ex Machina)

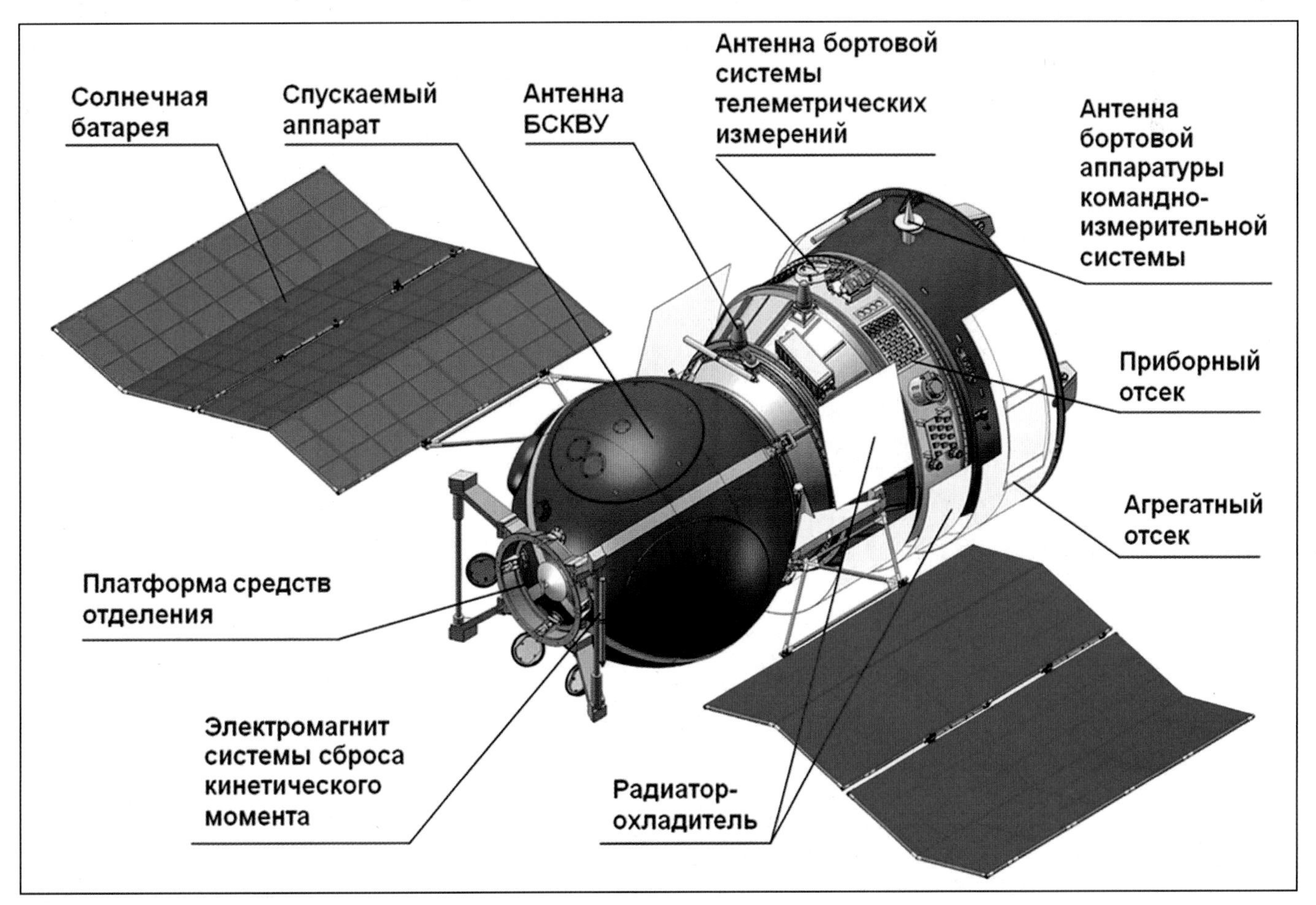

TODAY

SAVE UPTO £27 WHEN YOU SUBSCRIBE!

Aviation News is renowned for providing the best coverage of every branch of aviation.

shop.keypublishing.com/ansubs

Combat Aircraft Journal is renowned for being America's best-selling military aviation magazine.

shop.keypublishing.com/casubs

WEATHER WATCH

Remote sensing from space proves key to a global inventory of assets and capabilities.

ABOVE • Japan's Himawari Earth observation satellite provides a high-resolution image of the Far East. (JAXA)

BELOW • Harry Wexler became the first man to fly directly into a hurricane using a stripped-down Douglas A-20 Havoc of the type shown here. From that came his determination to produce the world's first weather satellites. (USAAF)

Satellites now provide information, data, and images of the Earth on a scale unimaginable when the first Landsat was launched in July 1972. It quickly became apparent that remote sensing from space was key to a global inventory of assets and capabilities – assets in the inventory of renewable and non-renewable resources and capabilities in terms of applying space applications to their more efficient and sustainable use to balance human and non-human life on an equal basis.

When Landsat began to provide added value to studies of the Earth, agriculture, geology and the rapidly changing impact of human expansion, fine-sounding ideals had been around for a decade or more. They would grow and embrace large sectors of the science community, blending environmental concerns with accelerated changes induced by human activity. These would lead to a connection between studies of the Earth's surface and indicators of unexpected change to the planet's climate. But there was another group of applications which did more than any other to link remote sensing to the changing climate – weather satellites.

Before NASA received the high-level mandate in May 1961 to put astronauts on the Moon by the end of the decade, the agency planned for a wide range of programmes covering every practical application of satellites and spacecraft, among which was a determination to discover just how practical it would be to use television cameras in space to send down pictures of weather systems in the atmosphere. This would give meteorologists a global view rather than stitching together information from isolated data-gathering spots around the planet.

In the past, the ability to precisely forecast weather conditions have frequently determined the outcome of battles and the winning of wars. To pursue their objectives, military strategists and tacticians are keen to understand conditions their personnel are required to operate in. It has been so since the earliest recorded days of mobilised conflict. Even before the American Civil War of 1861-1865, Prof Joseph Henry of the Smithsonian Institution set up a telegraph network which was used for meteorologists to send weather reports to the Union armies. During that conflict, balloons were employed to provide core information on wind, rain, and mist.

More recently, during the First World War, weather forecasters provided information on wind conditions to ensure that gas attacks would not blow the disabling toxin into the ranks of friendly troops. They frequently got it wrong. During that conflict, Norwegian scientists developed air mass measurements and techniques for measuring frontal systems which became core tools for understanding the weather and making reasonably accurate predictions.

The D-Day landings of Allied troops on the Normandy beaches on June 6, 1944, is a classic example of how weather forecasters provided pivotal information about when to launch a strategic offensive. Decisive information which delayed the landings for 24 hours to take advantage of a lapse in bad weather was relayed by Maureen Flavin Sweeney of the Blacksod Lighthouse in County Mayo, Ireland to Gp Capt James Stagg, chief meteorologist for the RAF, who made the final decision. Using a brief window of opportunity, the landing was a success.

Flight planning

At a more mundane level, during that conflict, the expansion of global logistical and freight movements by air required fairly reliable weather conditions to be provided to flight planners around the world. Large supplies were air-lifted from North America, across the North Atlantic to airfields in Scotland, or across the mid-Atlantic to Africa, where they were flown up across the Middle East and either to India or Russia. Other routes were more perilous, and meteorologists came into their own, especially when air freight moved across 'The Hump', over the Himalayan Mountains to China for attacks on Japan.

In the more general world of post-war civilian activity, accurate meteorological observations had successfully predicted intense storms, provided hazard warnings, and saved lives. In the commercial world, ordinary activities involving farmers and fishing fleets needed detailed weather predictions to make decisions about how to plan and organise the business of harvesting food and feeding an increasingly populous world. With the stimulus of the 1957-1958 International Geophysical Year, interest in obtaining accurate and predictable weather patterns grew.

It had been an enduring aspiration for meteorologists to obtain large-scale area coverage of the skies and to monitor, map and interpret weather patterns over great expanses of the atmosphere. The first successful V2 pictures of clouds from space occurred on March 7, 1947, when the US Army launched the German rocket to a height of 162.5km (101 miles) from White Sands, New Mexico. It raised interest around the world and weekly news magazines, so typical of the time, acclaimed the picture showed the Earth really was round, the curvature of the horizon clearly visible.

Two years later, in January 1949, US Air Force Major Dwain L Crowson published a paper discussing the possibility of using sounding rockets to take pictures of storms and to launch those frequently so as to build a sequence of images tracking their intensity and pace. On October 5, 1954, an Aerobee sounding rocket took a series of images which, when stitched together into a mosaic, showed a tropical storm for the first time across a significant expanse of the Earth's atmosphere. But it was a fixed-frame picture and was unable to show motion or swirl patterns to indicate intensity.

ABOVE • The general configuration of the Tiros satellite was built around a simple construction and with solar cells providing electrical power. (NASA)

LEFT • The underside of the Tiros satellite supported the optical equipment and the communications antennae for transmitting pictures to ground stations. (NASA)

At the dawn of the Space Age when NASA was formed, a lot of attention focused activity toward the benefits of space and the way it could be used to bring advantages to existing activity. To find new ways of doing ordinary things in a better and more efficient manner and to find new capabilities outside the existing inventories. Earth observation had been a key area of discussion but initially only the defence and intelligence community were involved in pushing that forward – in America with the Corona/Discovery programme and in Russia with Zenit spy satellites and associated programmes.

But civilian scientists wanted support for Earth observation at a more fundamental level and some people tried hard to push for an experimental weather satellite. One which could provide almost instant information because the weather has no pause button. It would only be of real use on a time-urgent basis so that forecasters could obtain direct observations of large areas and correlate data from balloons and static recording stations with large area pictures of cloud and perhaps detailed measurements of conditions within the atmosphere. The stimulus came from an unlikely source.

Birth of an idea

Born in 1911, Harry Wexler was a meteorologist with the US Weather Bureau and known for his derring-do. During the Second World War, he saw service and become the first person to deliberately fly into a hurricane. Equipping an adapted Douglas A-20 Havoc attack aircraft, he was on board to take measurements when it achieved this risky venture on September 14, 1944. After the war, he returned to the bureau and set his mind to studying the dynamics of the planet's atmosphere while also becoming interested in the atmospheres of other planets.

Wexler combined his interest in weather on this world and others with a fascination of planetary exploration and corresponded with Arthur C Clarke, the science

TOP • *NASA's William G Stroud (centre left) displays the inner workings of a Tiros satellite to Senator Lyndon B Johnson (right) in April 1960. (NASA)*

ABOVE • *Pictures were recorded on board Tiros using this magnetic tape data recorder. (Sanjay Acharya)*

RIGHT • *With weight at a premium, general configuration of the early Tiros satellites presented a masterclass in miniaturisation and compact design. (NASA)*

fiction writer and futurist, who wanted to know his views on a possible weather satellite. Clarke was a prolific communicator and had meetings with a range of mind shapers of his time. Included in these gatherings, frequently in a London pub, was Lord of the Rings author J R R Tolkien.

Inspired by Clarke's assertion regarding the wide range of possibilities for using satellites, Wexler began a campaign to get such a concept approved, proposing a World Weather Watch to create a global network of information from data-gathering platforms, including satellites connected to data relay stations for a truly global network of information which meteorologists could access.

Wexler was in a good position to get this discussed in a wider context. In 1954, he published a paper which mentioned the concept of a weather satellite for the first time. As chief of the Weather Bureau's Scientific Services branch, he was in charge of the US expedition to Antarctica for the International Geophysical Year of 1957-58 and knew a lot of key people. Never one to withdraw from establishing contacts and mobilising powerful lines of communication driven by conviction, Wexler gathered a consensus in support of his idea and attracted research at various laboratories in which the technical requirements for such a system were studied and discussed.

Wexler wanted to connect data gathering with new generations of computers and did much work on the concept of a 'nuclear winter'. In this, he proposed the theory that the planet could be plunged into a new Ice Age in the event of a global war with atomic weapons. And this was before the design and test of the infinitely more powerful thermonuclear (hydrogen) bomb. Wexler died of a heart attack in 1962 but lived long enough to see the world's first weather satellite in operation.

What he started began a series of developments in which NASA picked up the concept and attracted support from a wide range of interested government agencies, including the military. Initial studies had been conducted by the Advanced Research Projects Agency (ARPA) in 1958 before the project was handed to NASA on April 13, 1959, six months after the space agency had been set up. NASA defined those requirements, agreed the type of orbit which would be the best for providing such a service and outlined the design characteristics of the satellite itself. It had to fit within the size and weight constraints of available launch vehicles, which in the late 1950s were strictly limited.

Obvious restraint

With a specific set of objectives, the most obvious restraint was that of the Thor-Able launch vehicle selected to lift the satellites, which were to be known as TIROS, an acronym for Television Infrared Observation Satellite. The name defined characteristics of the satellite itself, which was to send television signals to Earth stations and provide images biased into the infra-red portion of the electro-magnetic spectrum. It was considered experimental, a test of ideas and concepts to see how the concept worked and to discover optimum operating modes for its future use.

The Thor-Able rocket was a derivative of the Thor, a ballistic missile developed by the US Air Force and adapted for a generation of satellite launchers. The Able upper stage had been designed as the second stage of the Vanguard rocket, which had been developed as America's satellite launcher. The optimum orbit for a weather satellite would have been in a polar or sun-synchronous path, allowing the same areas to be repeated on a daily basis, but the lift capacity of the Thor-Able prevented that. TIROS would be placed in an almost circular path of 693 x 750km (431 x 470 miles) in a path inclined 48.4° to the equator, covering a ground track 48° N to 48° S.

Initially, TIROS satellites would be small, weighing just 122.5kg (270lb). Each consisted of a drum-shaped structure

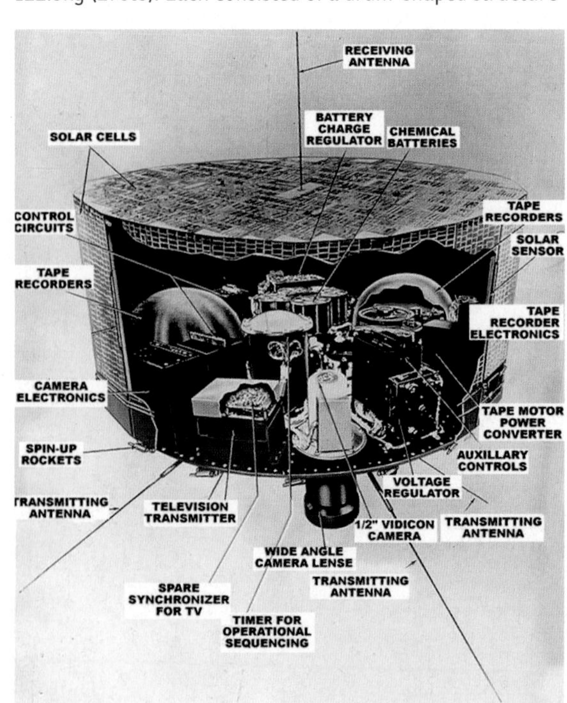

with a diameter of 110cm (42in) and a height of 56cm (22in) but faceted with 18 flats carrying 9,000 individual solar cells providing electric power to 21 Ni-cad batteries. Small thrusters would maintain a spin rate of 8-12rpm for gyroscopic stability with two small 1.27cm (0.5in) diameter vidicon cameras for taking wide-angle and narrow-angle images which would be transmitted directly to the ground when the satellite was over a receiving station or stored in on-board recorders and transmitted later.

The application of the video tube allowed useful images to be obtained without the bulk of a standard television camera. Based on the cathode ray tube (CRT), the image would be scanned by an electron beam which generated an electrical current proportional to the brightness and a picture formed from a series of scanned lines. In early US television sets, the standard images were formed of around 472 lines in NTSC format or 576 lines in the European PAL system, up to 1,035 lines being provided in high-definition sets using the old CRT system.

NASA took advantage of work conducted by RCA into a satellite design after this organisation had responded to interest during the time ARPA was looking into TV-based weather satellites. NASA would contract RCA to continue this work and to provide the cameras. The requirement was for a wide-angle camera to take images of large cloud structures in pictures showing an area 1,125km (700 miles) wide while the narrow-angle camera with images 101km (63 miles) wide would help identify different cloud types seen in the broader image. The wide-angle camera had a resolution of 5km (3.1 miles) and the narrow angle could show cloud structures 0.5km (0.3 miles) across, quite adequate for the job.

Images would be stored on two magnetic tape recorders, each with two separate channels for the wide and narrow angle pictures. Designed and assembled by RCA, they brought new and demanding techniques to manufacture, test and flight operations, giving the company experience which was expressed through other programmes with similar demands on size and minimal weight. Each recorder carried 122m (400ft) of 0.24cm (0.094in) tape for 32 images and would operate at a speed of 127cm (50in) per second. Pictures would be reproduced on 500-line TV screens.

Cloud coverage

The object of TIROS was to provide large-area coverage of cloud activity which was not required to have a surface resolution to show features and ground structures. Early discussions by the defence and intelligence community on whether to use digital transmission or wet-film recovery for spy satellites favoured ground processing for higher surface resolution pictures in these early days of digital transmission. Hence the decision to develop Corona satellites with recoverable film. For weather forecasting, the cloud cover pictures from TIROS were adequate.

The value of such a system was in the global context by which the information was obtained and applied. With conventional data-gathering tools, large areas of the world's weather systems in northern and extreme southern latitudes were void of any reliable information. While TIROS was experimental and could not cover these polar regions, the expectation for expansion into a truly global net was the aspiration that powered early development. Moreover, many areas of the world had no monitoring capability at all, and atmospheric conditions and cloud cover data obtained on a daily basis over large areas of the ocean were totally unknown.

TIROS flies

The world's first weather satellite, TIROS I, was launched in the early dawn of April 1, 1960, with the Thor-Able rocket performing as expected. The first two stages of the rocket placed the third stage and the satellite on course for orbital insertion as the payload crossed over the west coast of Africa. The first pass over the tracking station at Woomera, Australia, showed that all was working well and that the spin rate was around 10rpm – exactly as expected. On across the Pacific Ocean it passed before transiting North America, its first pictures being taken over the Gulf of St Lawrence and out over the Princeton, New Jersey tracking station.

BELOW LEFT • *Engineers attach a Tiros satellite to the Able upper stage which will be carried on top of a Thor launch vehicle. (NASA)*

BELOW • *The world's first bespoke weather satellite, Tiros 1, lifts off from Cape Canaveral on a Thor-Able launch vehicle at 6:40am local time on April 1, 1960. (NASA)*

RIGHT • *The Tiros control centre, where command of the satellites took place and to where the images would be collected and analysed before being passed along for interpretation. (NASA)*

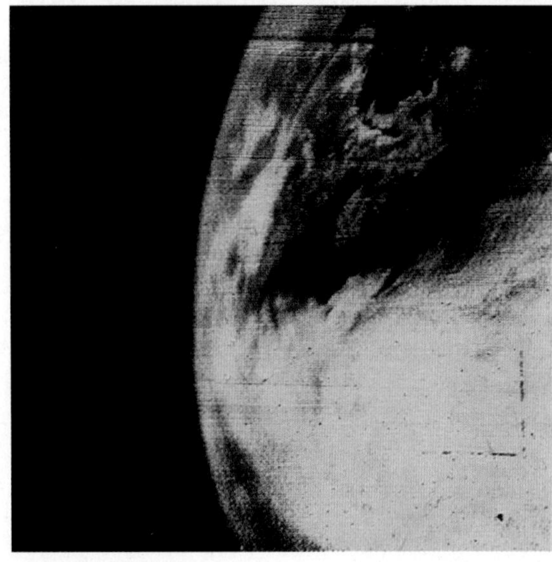

RIGHT • *The first picture returned from space showed cloud patterns and weather systems but little else. (NASA)*

Having sent the last of 22,952 frames of weather systems on June 15, 1960, this first satellite of its kind operated for a further two weeks. In 1,303 orbits of the Earth, it had more than proved its worth, for its time a highly successful experiment in which both the launch vehicle and the satellite had performed well. For its time, where more than half the rockets and payloads launched were failures, that in itself was high success and led to great confidence as preparations were finalised for the launch of TIROS II.

That occurred on November 23, 1960 and included the infra-red vidicon tubes which had not been ready for the first flight, TIROS I carrying only tubes for the visible portion of the spectrum. But TIROS II had problems, only 10 per cent of the wide-angle pictures were usable, although 85 per cent of the narrow-angle images were good. But the satellite sent down more than 36,000 pictures and the entire system proved useful in providing operating experience. There was much still to learn about data transmission, handling, and image interpretation.

Launched on July 1, 1961, TIROS III carried an advanced radiation experiment which replaced the infra-red system. More than 100 foreign weather stations were recruited to help share the images and to correlate their own data gathering from more conventional land, sea, and air instruments with the satellites from space. An outstanding success, TIROS III alerted weather watchers to Hurricane Esther two days before the airborne hurricane hunters could provide this information. A total of 75 weather alerts were issued by the Weather Bureau as the satellite reported five hurricanes and a tropical storm in the Atlantic, two hurricanes and a tropical storm in the eastern Pacific and nine typhoons in the central and western Pacific. By end of 1961, the mission was over.

Despite being hailed an experimental system, the weather satellites kept on displaying a semi-operational capability, with TIROS IV sent up on February 8, 1962, from which the Weather Bureau began to issue daily international transmissions of cloud cover maps based on its photography and encouraging international participation. Much of this had been stimulated by a conference held on November 13-21, 1961, hosted in Washington DC, by NASA and the Weather Bureau, where representatives from 28 countries attended to work out how best to incorporate TIROS data into their own national weather forecasting activities.

Transition process

To begin the process of transition to an operational system, TIROS V was launched on June 19, 1962, in time for that year's hurricane season. This satellite did not carry infra-red and to obtain images of regions at higher latitude, it was delivered to an inclination of 59° so that it could obtain

ABOVE • *Tiros-M was renamed ITOS (Improved Tiros) with the second in the series becoming NOAA-1 when administration of weather satellites was taken over by that organisation. (NOAA)*

periodic information from near-polar regions. TIROS VI went up on September 18, 1962, two months earlier than scheduled so as to add coverage for the hurricane season, both to assist with its predecessor and to provide extended weather coverage for the Mercury flight carrying astronaut Walter M Schirra.

TIROS VI had begun the process of having two weather satellites operational at the same time in preparation for the more advanced Nimbus weather satellite. TIROS VII was launched on June 19, 1963, followed by the last of the series on December 21, 1963, being the first to carry a new experimental camera known as the Automatic Picture Transmission (APT) system. APT allowed direct readout of weather pictures to users all over the world and by the end of the year fifteen international stations were ready to receive pictures.

By the end of 1963, the eight TIROS satellites had transmitted 239,360 usable images for 8,199 cloud cover analyses, producing 1,253 storm advisories with 18 hurricanes and 32 typhoons analysed and tracked. Every launch had been a success and all satellites had operated, not always to the technical standards required but none had failed in their primary mission. The Russians had labelled TIROS as a spy satellite and denounced it as an aggressive and hostile act but when Presidents Kennedy and Khrushchev exchanged messages between each other, it was agreed that both countries would share meteorological data in the common interest, a document to that effect being signed in April 1962.

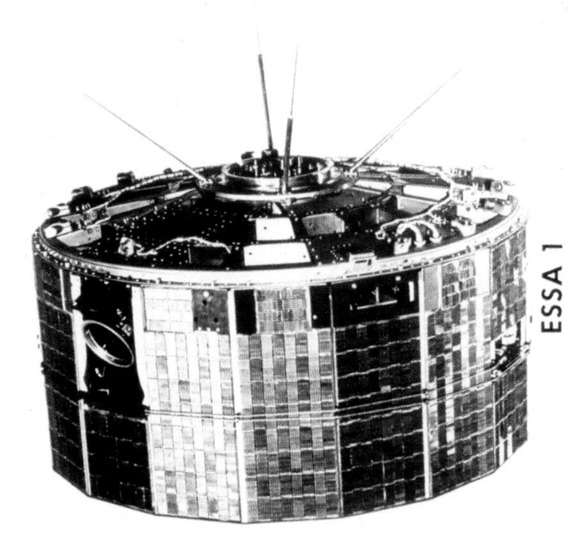

LEFT • *Precursor to NOAA, the ESSA series of weather satellites was based on the Tiros design and incorporated a side-looking camera instead of the vertical camera aligned with the spin-axis. (NOAA)*

More was to come, and plans existed to launch the new Nimbus satellite in 1964, a year in which, on March 20, NASA and the Weather Bureau signed an agreement for development of TOS – TIROS Operational Satellite system. This would have a very different design and provide expanded capabilities. All would carry the APT system. The Weather Bureau agreed to finance TOS and NASA would launch them and operate the ground stations. It would be connected to the United Nations' World Meteorological Organization.

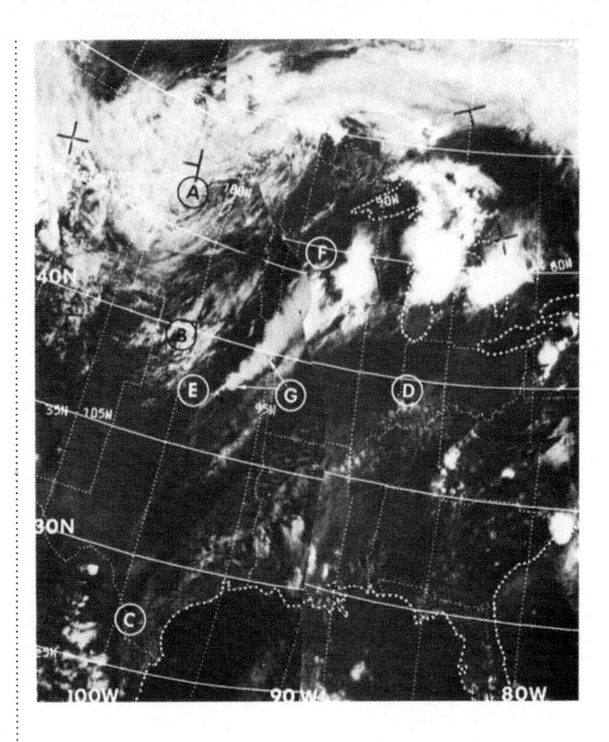

RIGHT • A photo-mosaic of images from an ESSA satellite over the United States as weather pictures from space assumed a more operational role in national meteorological services. (NOAA)

Harry Wexler lived to see the first of the series launched and had the following comments: "We are now able to study the mysterious cloud formations and movements, viewing them in pictures taken from many miles above. In Earth-orbiting satellites, the meteorologist now has for the first time an observing platform that increases the range of sight from a visual 20 miles and from radar's 200 miles to a truly global scope."

The bigger picture

The TIROS Operational Satellite programme was not the definitive, space-based weather system implied by its title. There was still much to learn about the best services to provide and the optimum technology to employ. Moreover, with the range of launch vehicles becoming available in the mid-1960s, heavier and more capable satellites could be planned, and the satellites would have to operate from a polar orbit to gain the best value and ensure optimum coverage for the global user community now coalescing around the programme.

In the overall structure of government affairs, weather satellites experienced a soaring ascent to the top table. They inspired the formation of the Environmental Science Services Administration (ESSA), which was to manage US meteorological, climate, sea and ocean studies and the general mapping of the Earth's structural shape. It would be superseded by the National Oceanic and Atmospheric Administration in 1970. ESSA gave its name to a group of satellites designed to provide transition from an experimental to a fully operational space-based weather service.

Behind the TIROS and ESSA satellites were the Nimbus technology development platforms for testing advanced systems. Each was launched by a Thor-Agena or Delta rocket to a sun-synchronous orbit from Vandenberg Air Force Base on the coast of California. From the first sent aloft on August 28, 1964, to the last on October 24, 1978, the seven Nimbus satellites pioneered new ways of observing the Earth and served as a precursor to the Landsat series first launched in July 1972.

Each satellite was built by General Electric and RCA and had the same appearance as the Landsat satellite upon which that platform was based. Nimbus 1 weighed 374kg (825lb) but as instruments were added and the configuration grew, weights would increase to 965kg (2,127lb) for the last. The second spacecraft was lost when it suffered a launch failure on May 18, 1968. It had been equipped with a Snap-19 radio-isotope thermos-electric generator (RTG) nuclear power source which fell to the

BELOW • The ESSA-4 satellite gets away from Cape Canaveral on January 21, 1967, atop a Thor-Delta rocket with three solid propellant, strap-on boosters. (NASA)

floor of the Pacific Ocean. This was the only failure in the Nimbus programme.

Each satellite carried a unique arrangement of instruments for observing the Earth, including cameras with various filters, radiometers, spectrometers, and scanners, together with beacon transmitters. With enhanced capabilities, Nimbus also opened the door to a global position system (GPS) for navigation. Satellites for global navigation emerged in the 1970s and Nimbus also demonstrated how satellites could be part of a global search and rescue network. British aviator Sheila Scott made the first test of the satellite's navigation and locator communication system as she made the world's first solo flight over the North Pole in 1971.

They served the meteorological community by making three- to five-day forecasts routine, collecting ocean and air temperatures, air pressure measurements and cloud levels by density and mass. They also measured the difference between the water vapour in the atmosphere and the amount of liquid water in clouds. These significantly extended the scientific understanding of the atmosphere and played a major role in measuring the ozone layer and mapping its depletion. They did pioneering work on sea ice mapping and measurement of ice floes, glaciers and melt zones, providing the first global data on declining levels of global cover.

Nimbus also measured the amount of solar radiation entering the atmosphere and the quantity being reflected back. That provided the basis for understanding how the Earth was inhaling and exhaling and how significant changes were occurring in that balance, with a net increase absorbed and more energy building in the atmosphere. In turn, during the early 1970s, this led to the first direct measurement of a changing climate and the initial calculation about significant effects from an increase in the average temperature of the planet.

Trigger point postulation

It was from this information that a broadly accepted figure of 1.5° C increase was postulated as the trigger point for irreversible change affecting the atmosphere and most living organisms on the planet, a level generally quoted today. By the mid-1970s, TIROS, Nimbus and Landsat satellites had built the bedrock on which global environmental science was based. Unexpected, and with some surprise at the data, scientific opinion was solidly locked to warnings of potential catastrophe due to human activity, with the loss of many species and an accelerating impact on society at large.

Meanwhile, NASA and the Weather Bureau began the transition to the TIROS Operational Satellite system with the launch of TIROS IX on January 22, 1965 and TIROS X, on July 2, both in time for the hurricane season that year. Precursors to the operational system to come, instead of inclined orbits at about 48-58° to the equator, they both went into sun-synchronous paths of 96.4° and 98.6°, respectively. Weighing in at 138kg (304lb), TIROS IX had its two cameras mounted on opposite sides of the drum rather than in the base plate parallel to the spin axis as configured for all previous satellites and for TIROS X.

A magnetic attitude control system aligned the spin axis perpendicular to the orbit plane and at a tangent to the Earth's surface. This orientation and control was adopted for the operational satellites to follow and the satellite completed the first photo-mosaic of the entire world's cloud cover through a composite of 450 separate photographs taken on February 13, 1965. For the first time, it was possible to see the planet's weather in its entirety. The TIROS programme had started as an experimental platform for testing an idea which has grown rapidly into a reliable and impressive system, drawing in a global user community.

Management and operation of the next generation TOS series was in the hands of the Environmental Science Services Administration. NASA had done its job of crafting a workable system with effective instruments and operationally reliable satellites. Accordingly, the TOS series of satellite were designated ESSA-1 through to 9, launched between February 2, 1966 and February 26, 1969. They were all successful, with the last (ESSA-8) retired on March 12, 1976. All were launched into

LEFT • *Concurrent with the Landsat remote-sensing platform, the Nimbus series of satellites was advanced and experimental in practice, each providing slightly different payloads and with a general trend towards weight growth across the seven launched between 1964 and 1978. (NOAA)*

BELOW • *The advanced Tiros-N series of the polar-orbiting weather satellites weighed 2,232kg (4,921lb) and carried a single solar array. (Lockheed Martin)*

sun-synchronous orbits by Delta rockets from Vandenberg Air Force Base and all were capable of carrying a range of sensors, camera systems and other instruments.

Continuation of the civilian weather satellite programmes saw the first of the Improved TIROS (ITOS) family, which was subjected to a confusing set of names and identifying numbers which reflected the primary change of ownership and operating control. The National Oceanic and Atmospheric Administration (NOAA) replaced ESSA in October 1970 and satellites which could just as easily be identified as TIROS, ITOS or ESSA assumed the badge of NOAA, by which most of the ITOS series are known for the duration of the 1970s, when six were launched between January 23, 1970 and July 29, 1976, the first of which was named TIROS-M. Two of these failed to reach orbit.

Many and varied

While the United States had an early start on weather satellites, the Soviet Union did not launch the first of their Meteor-1 satellites until March 26, 1969. With a weight of 1,200kg (2,646lb), it operated from an orbit of 650km (404 miles) and an inclination of 81.2°, which was characteristic of all 27 such satellites in this series sent aloft by April 5, 1977. The Meteor-1 series were capable of vertical profiling of temperature and moisture levels in addition to observations of global weather, cloud patterns and surface ice and snow fields.

This series was followed by Meteor-2 types, the first launched on November 7, 1975, with three TV cameras in the visible and infrared parts of the spectrum, a five-channel radiometer, and an instrument for measuring changes in the radiation levels at the outer part of the Earth's atmosphere. The satellites weighed 1,300kg (2,866lb) and were of essentially the same design as the Meteor-1 series with two deployable solar wing panels. The last of 21 was launched on August 31, 1993.

Russia's Meteor-3 weather satellites appeared with the launch of the first on November 27, 1984. With a weight of up to 2,250kg (4,961lb), these were semi-experimental and partly operational in that the six launches carried different instruments, more advanced than those on earlier weather satellites and with greater input from a wider range of government departments. A derivative development, the 2,477kg (5,462lb) Meteor-3M was planned as a series of four polar orbiting satellites with a visible channel providing 1.4km (0.87 miles) resolution and a 10-channel radiometer with a resolution of 3km (1.8 miles). Only one was launched, on December 10, 2001.

A major programme designated Meteor-M evolved after the collapse of the Soviet Union, with a satellite weighing 2,750kg (6,063lb), providing low resolution multi-spectral scanning for cloud cover mapping, instruments for obtaining atmospheric temperature and humidity profiles and for measuring sea surface winds. There was a capability for advanced infra-red soundings of the atmosphere providing vertical profiling to show various conditions at different altitudes and special data collection and integration equipment.

The first was launched on November 28, 2017, but the rocket failed, and it never reached orbit, the second on July 5, 2019 being a success until a malfunction caused by a debris collision in orbit on December 18 ended its operation. More were planned but none have been launched so far.

On January 2, 2011, Russia launched the first of four Electro-L weather satellites to geostationary orbit with the purpose of providing visible and infra-red images showing the full disc of the Earth with a fresh image every 30 minutes. With a weight of 1,740kg (3,837lb), the satellite was initially placed over the Indian Ocean at 76° E longitude. Its propulsion system raised the orbit by 289km (179 miles) on July 13, 2016, which unlocked the period of the orbit matched to the rotation period of the Earth. Moving slightly faster than the Earth's speed of rotation at this slightly higher altitude, it began to drift westward until it reached its final position at 14.5° W by October 3, at which point the altitude was lowered to a geostationary position once more. Three more were launched to geostationary orbit, the most recent on February 7, 2023.

Geostationary orbits attracted the first European weather satellite under the Meteosat programme, which was approved in 1972 and saw launch of the first in a series of seven on November 23, 1977, by Ariane rocket,

*BELOW • **NOAA-19 suffered severe damage when it fell over during final assembly in 2003. Repaired, it was launched on February 8, 2009. (NOAA)***

the last going up on September 3, 1997. With a launch weight of 696kg (1,535lb), each satellite consisted of a drum-shaped structure 2.1m (6.9ft) in diameter and with a total length including antennae of 3.1m (10.4ft), covered with 8,000 solar cells producing 200W of electrical power and spin-stabilised at 100rpm. On station positioned over Africa, it provided visible and infra-red images every 30 minutes in addition to measuring water vapour in the atmosphere.

Growing awareness

With a growing global awareness of the increasing demand for weather data from countries around the world, several organisations were set up to handle national, regional, and global exchange of information, freely and for mutual benefit. A consortium of several European countries came together in 1986 to form the European Organisation for the Exploitation of Meteorological Satellites (EUMETSAT) and which today represents 30 countries across the continent.

ABOVE • Designated NOAA-20 and partnered in sun-synchronous orbit with the Suomi satellite, the 2,540kg (5,600lb) spacecraft is part of the Joint Polar Satellite Program (JPSP). (NOAA)

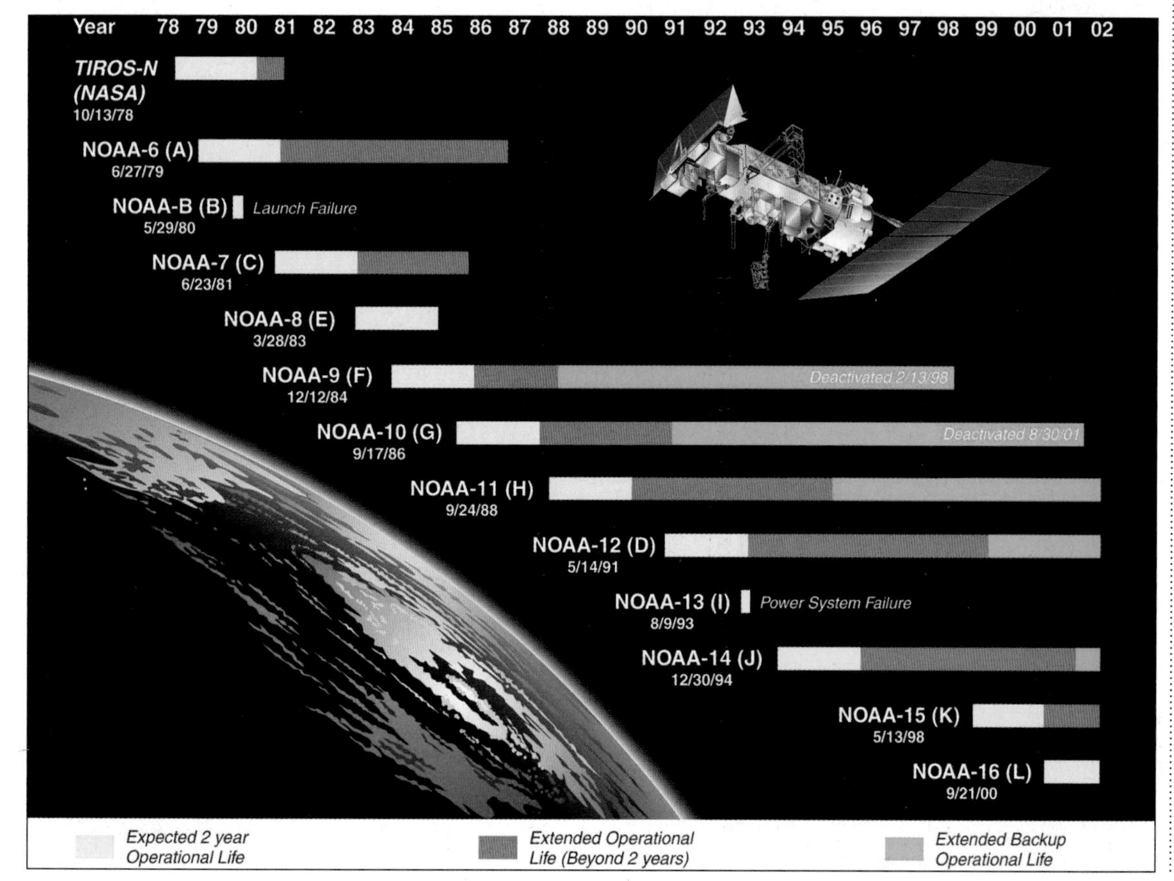

LEFT • A time chart showing the later generations of Tiros-N weather satellites as the programme evolved to the end of the 20th century. (NASA)

ABOVE • *Europe's third-generation geostationary Meteosat programme provides national, regional, and international support for the global watch on weather. (ESA)*

National weather forecasting bodies, and an increasing number of commercial organisations providing meteorological data, get their information from these massive data banks. Equipped with some of the world's most powerful computer processing systems handling terabytes of data each day, the worldwide system provides the same level of information for Sunderland and Singapore as it does for Samoa and Senegal. With these highly sophisticated systems, information flows continuously as Earth's weather is monitored on a sustained and unrelenting basis.

Europe, the United States, Russia, India, and China all pool data from national weather satellite systems to multiply the value of that far beyond the level of data that could be supplied by a single organisation or country. Integration of a global interdependence on information about the planet's weather systems began to empower a more general survey, integrated with environmental satellites, of the dramatic changes to near-term climatic events as violent storms, unusual rainfall, excessive and enduring heat, and eroding ice shelves. What had begun as a desire to improve weather forecasting became a visual diary of a chaotic atmosphere.

With a lead taken by the United States and followed by Europe, the incentive to develop new and more powerful diagnostic tools grew. A lead was seized with Europe's successors to Meteosat's first generation platforms. Four satellites, designated Meteosat-8 to -11 were second generation designs, each weighing 2,040kg (4,498lb) and launched by the powerful Ariane V. Each satellite consisted of a cylindrical drum-shaped structure with a diameter of 3.2m (10.5ft) and a height of 2.4m (7.87ft), the complete assembly standing 3.7m (12.1ft) tall. The circumference of the main body of the satellite was covered with 7,854 solar cells providing 750W of electrical power.

Transition time

Meteosat is notable for marking the transition from a weather surveillance system to a satellite optimised

RIGHT • *Europe's Meteosat Second Generation (MSG) satellites are operated by EUMETSAT, which embraces 30 countries with data from the geostationary satellites, this image from MSG-4 from its position above the equator. (EUMETSAT)*

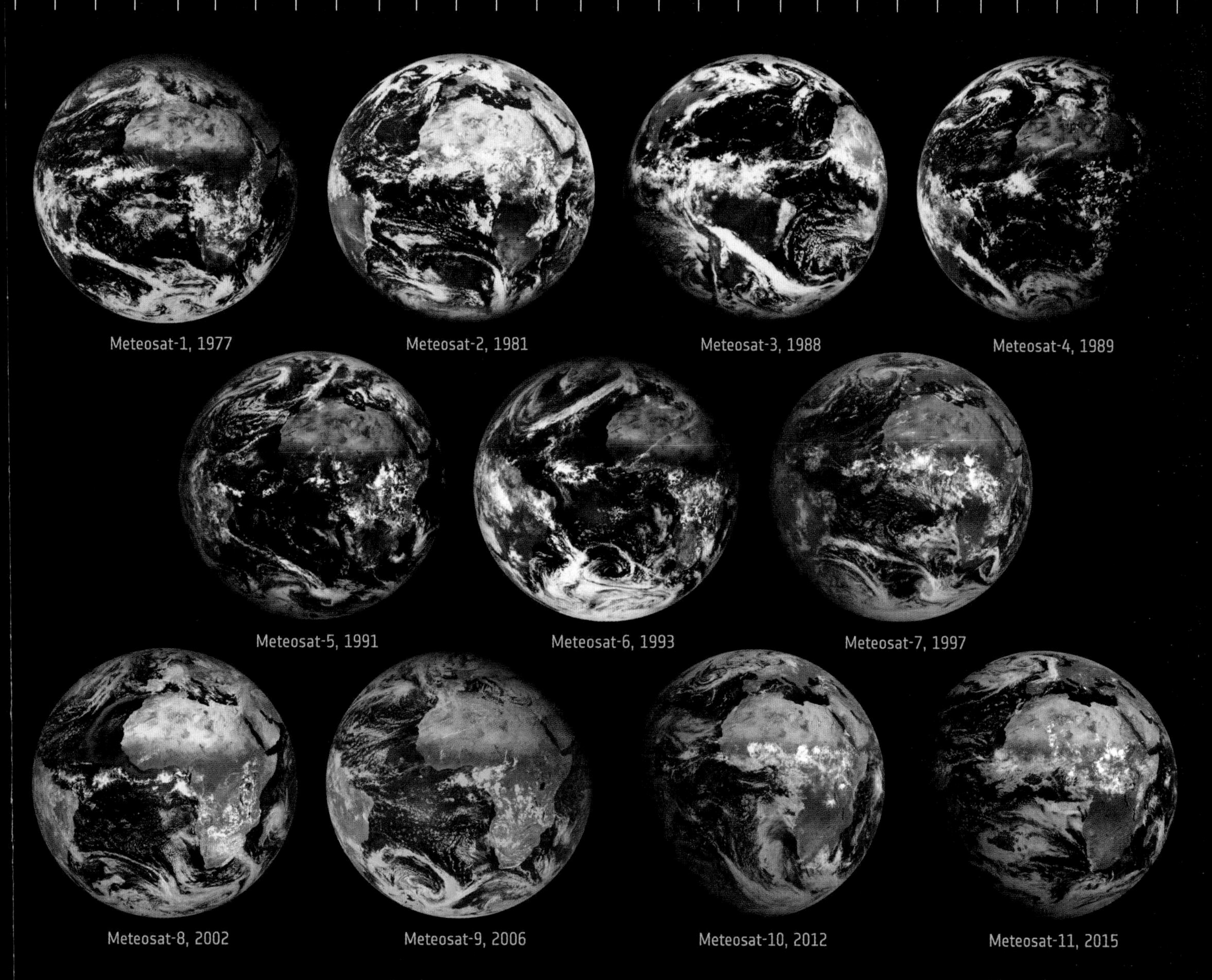

ABOVE • The evolution of Europe's Meteosat programme has been a major contributor to enhanced global weather information management from 1977 to the present. (ESA)

for both meteorological and climate research, a field of study that was becoming increasingly important when the programme was developed in the mid to late 1990s. Each of the four satellites carried a Spinning Enhanced Visible and Infra-red Imager (SEVIRI) and a Geostationary Earth Radiation Budget (GERB) instrument observing in two spectral bands. Both captured an image every 15 minutes for transmission direct to the ground at a pixel resolution of about 44.6km x 39.3km (27.7 x 24.4 miles).

The plan was to have two satellites in geostationary orbit, one acting as a back-up, with ESA responsible for the design and launch of the satellites and EUMETSAT for operations. The first (Meteosat-8) was launched on August 28, 2002, followed by the second on December 22, 2005, the third on July 5, 2012 and the fourth on July 15, 2021. The operational performance demonstrated its ability to provide 'nowcast' forecasting, data that is a mere 2-5 hours old for immediate computer interpretation into a valid projection of weather. The GERB instrument proved particularly valuable in measuring parameters vital for climate modelling.

ESA and EUMETSAT took the results, the data and the user requirements expressed by international bodies to plan for a third generation Meteosat system. Each satellite weighs 3,760kg (8,290lb) and, unlike all previous European weather satellites in this series, they are stabilised in attitude and not spinning. They are the most advanced geostationary satellites ever launched and provide images every 10 minutes from an instrument staring at the atmosphere through 16 spectral channels. A special lightning imager observes flashes between clouds and lightning strikes to the ground with high resolution imagery capable of tracking rapidly developing storm systems and providing early alerts of dangerous weather systems. Designated MTG-I1 (Meteosat-12), the first was launched on December 13, 2022 by Europe's Ariane rocket.

With a strong demand for data about the polar regions, ESA, EUMETSAT and NOAA co-operated on the development of three satellites known as MeTop (Meteorological Operational Satellite). Each weighing 4,093kg (9,025lb), they would measure temperature, humidity and wind speed and direction over the oceans and provide coverage of land and sea for general weather observations. Launched in 2006, 2012 and 2018, the satellites were designed to last about five or six years but far outlasted that prediction. MeTop-A was retired in November 2021 and the other two are still operating, data acquired supporting US polar research weather satellites.

Many other countries contributed to the World Weather Watch programme of the World Meteorological Organisation, notable being the Himawari geostationary satellites developed by Japan and first placed in orbit in July 1977. The most recent, Himawari 9, was launched in November 2016 and follows its predecessor operational since 2015. Both continue in service.

CARBON DIOXIDE OVER 800,000 YEARS

carbon dioxide (ppm)
450
400
350
300
250
200
150
100
current concentration
2021 average (414.7 ppm)
highest previous
concentration (300 ppm)
warm period
(interglacial)
ice age
(glacial)
800,000
600,000
400,000
200,000
0
years before present
NOAA Climate.gov
Data: NCEI

ABOVE • The climate dashboard of atmospheric carbon dioxide levels over the last 800,000 years with the current level in 2021 far higher than anything recorded since the last ice age. (NOAA)

RIGHT • Observation of rapid change in surface conditions by remote-sensing satellites tracked the evaporation of the Aral Sea as seen here in 2021. (NOAA)

now 30 per cent higher than in pre-industrial times. Bad news for fish and all marine life, so much of which cleans the oceans.

Analysis of global measurements that began with the 1957-58 International Geophysical Year (IGY) helped clarify the importance of carbon dioxide to the atmosphere: a little too much and it causes the greenhouse effect; too little and the Earth freezes over because not enough heat would be absorbed to distribute thermal energy from the day to the night side of the planet. The IGY that stimulated the world's first satellites provided clear evidence that changes were occurring at a rapidly increasing rate. But just by how much and at what rate?

In the years of the IGY, measuring stations on the Mauna Loa Volcanic Observatory, Hawaii, began recording global carbon dioxide levels, at that time 315ppm. Today levels are higher than they have been for three million years. At that time, with density levels at their current values, the temperature of the atmosphere was up to 4° C warmer and sea levels up to 25m (82ft) higher than they were in 1900. Using these historic figures and with detailed satellite-based data recovery, at current rates of increased energy demand, carbon dioxide levels could reach 800ppm within 75 years.

Mobilisation

Concern about the broader implications of data from weather satellites spurred an interest in developing a range of satellite programmes for detailed study of the Earth's climate. The word 'environment' crept into terminology for reorganising the structures of government to formally recognise the need for more information on the Earth and the atmosphere over longer time scales and across wider zones of scientific interest.

Of course, the study of the environment had been a key application for Landsat satellites and the many other programme run by NASA, ESSA, and NOAA. But there was a new urgency to focus now on the way the planet was changing globally, to study its oceans specifically and to link these to concerns about how the planet could adapt to these human-induced changes. NASA and the environmental agencies had developed a series of satellites increasingly tuned to obtaining scientific data, but data came not only from the civilian programmes. The military too had specific requirements for weather satellites which had been satisfied by a dedicated programme run by the Air Force.

Under the US Defense Meteorological Support Program (DMSP), on July 19, 1961, the Air Force approved a series of weather satellites specifically tuned to operational needs of the military. There was a requirement to obtain cloud cover

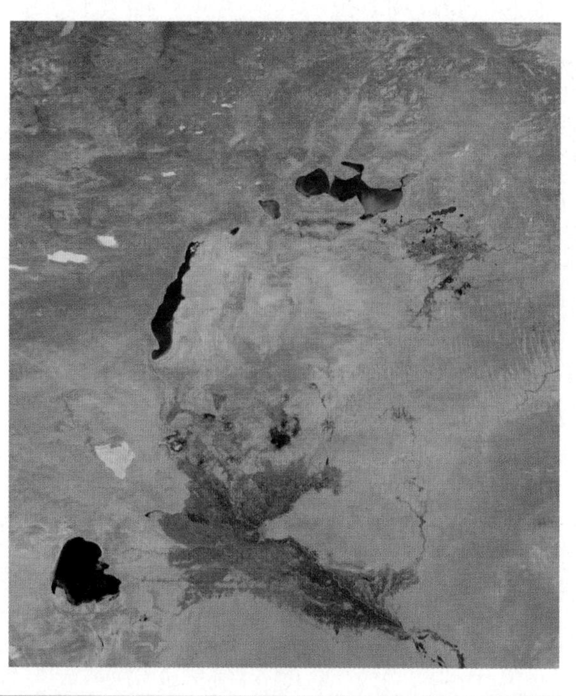

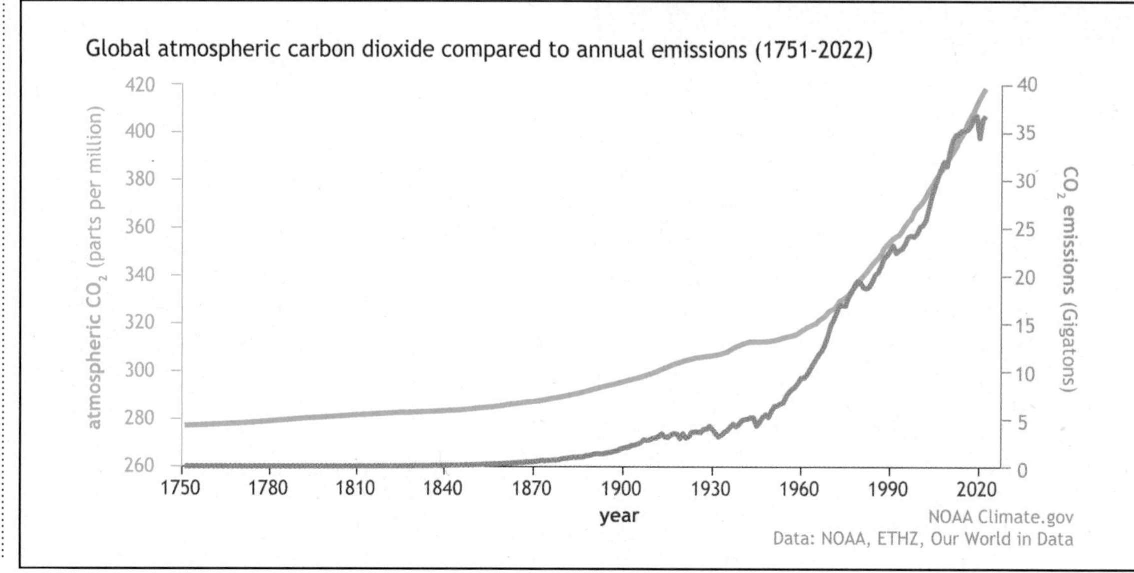

RIGHT • Ground-based observations between 1751 and the Space Age focus attention on very recent changes, with climate satellites recording levels over the last 50 years. (NOAA)

images of areas the emerging spy satellites would be required to photograph, rendered more efficient if they could avoid wasting film on cloud-shrouded targets. Information would be directly transmitted to two receiving stations in the United States, at Fairchild AFB, Washington and Loring AFB, Maine. Images, in both visible and infra-red, would be sent to the Air Force Global Weather Central at Offutt AFB, Nebraska.

Built by RCA, DMSP satellites operated in sun-synchronous orbits with a period of 101 minutes, completing 14.3 orbits of the Earth in 24 hours, which provided imaging opportunities over the entire surface twice a day. Operations were highly classified and even the origin of the programme had been veiled in excessive secrecy, Air Force chiefs believing that with the government heavily financing the NASA TIROS programme, there would be objections to a separate weather service for the military. The real reason for secrecy was the special observations of spy satellite targets.

The first successful launch occurred on August 23, 1962, using a Scout launch vehicle and a series of initial verification flights were completed on December 18, 1963. The first fully operational Block I satellites were launched from January 19, 1964, with the last of nine on May 20, 1965. For these and subsequent DMSP satellites, larger launch vehicles were employed to provide a fully operational service. Gradually, throughout the 1960s and early 1970s, the satellites transitioned from a strategic to a tactical role and were used extensively during the Vietnam War for supplying land, sea, and air units with photographic weather updates.

Over time, the weight of DMSP satellites increased to around 193kg (425lb) and more powerful launch systems were required. Known as the Block 5D-1, from the early 1970s a much heavier and more capable satellite appeared with a weight of 522kg (1,150lb) and a completely different design more like the advanced, three-axis stabilised TIROS satellites. The first 5D-1 was launched on September 11, 1976 and more launches followed as further improvements were introduced over the next decade.

In 1994 NOAA, the Department of Defense and NASA converged their respective programmes for weather and environmental research into a single effort known as the National Satellite Polar Orbital Environmental Satellite System (NPOESS). It was a direct response to a need for integrated data gathering and environmental research for all US government agencies. But it had its flaws.

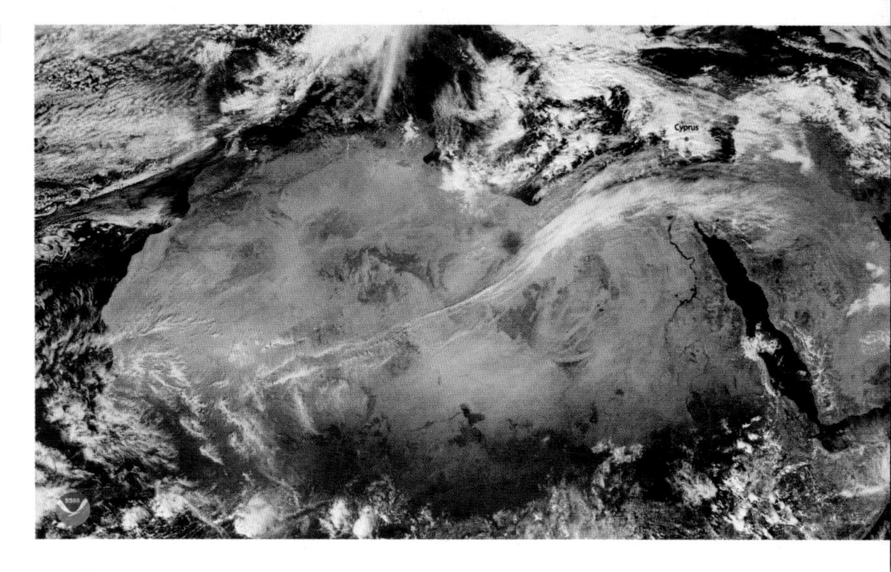

ABOVE • ***Westerly trade winds blow across North Africa from Saudi Arabia to Morocco affecting both the weather and the local environmental conditions as tracked and recorded by satellite. (ESA)***

In 2010, it was replaced by two initiatives: the Joint Polar Satellite System (JPSS) run by NASA and the NOAA and the Defense Weather Satellite System (DWSS), which evolved into the Weather Satellite Follow-On Microwave (WSF-M) two years later. That has produced a satellite scheduled to fly in 2024 to a sun-synchronous orbit carrying a microwave imaging radiometer for measuring ocean surface winds and charged particles which affect low orbiting satellites.

In a broader context, scientists are interested to measure the amount of ice in existing fields, flows, glaciers and on high elevation tundra and mountain ranges. Not directly associated with carbon measurements, the effect of increased heat loads into the atmosphere raise the temperature and melt ice. In this regard, satellites that measure ice in all these environments help test the effect of carbon dioxide emissions on how those enhanced temperatures are influencing ice density and melt rates across the globe. To do that, NASA launched the 970kg

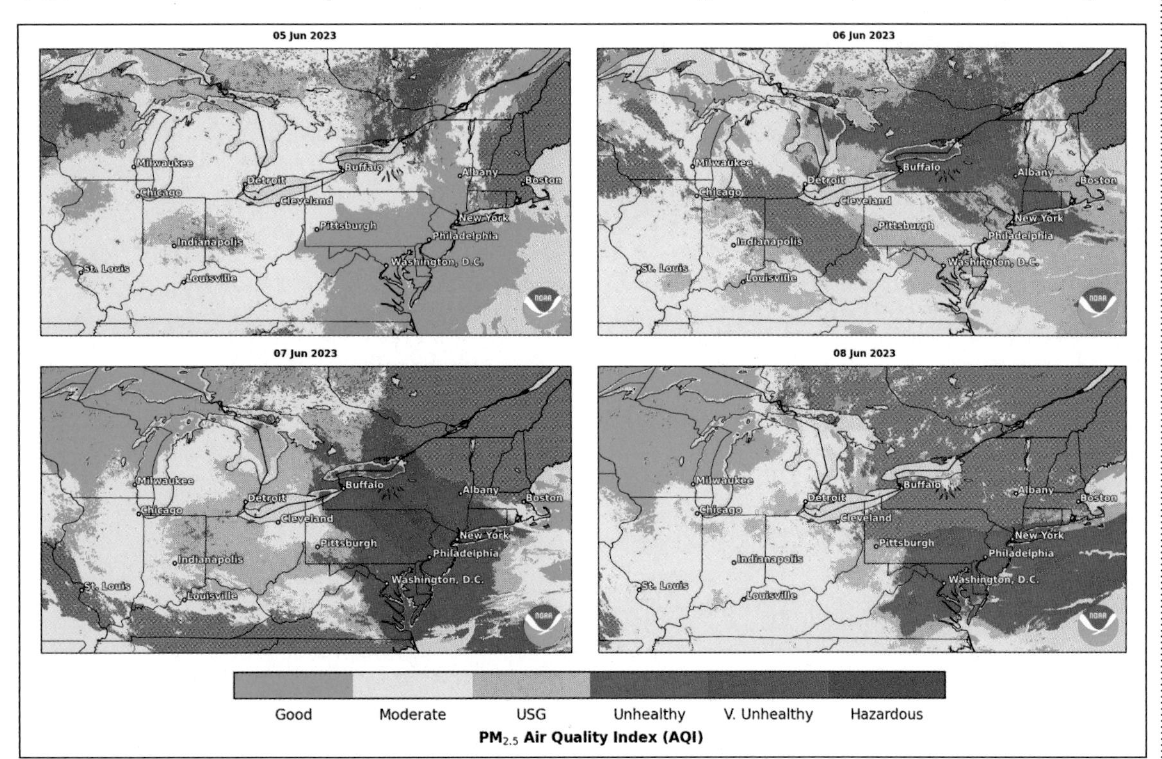

LEFT • ***Satellites can monitor the air quality index, deterioration on a rapid scale revealed here over a three-day period across the northeastern United States in June 2023. (NOAA)***

RIGHT • *Military weather satellites of the Defense Meteorology Satellite Program (DMSP) have evolved in parallel with civilian platforms, contributing a rich harvest of data used by scientists to study the changing climate. (USAF)*

RIGHT • *A military DMSP satellite records the northern lights over Norway as its instruments study the interaction of the Earth's weather systems with the near-space environment and the solar wind from the Sun. (USAF)*

(2,140lb) ICESat (Ice, Cloud, and land Elevation Satellite) to a near-polar orbit on January 13, 2003.

ICESat-2 followed with a launch on September 15, 2018. Weighing 1,514kg (3,338lb), it was specifically configured to measure ice sheet mass and to provide a topographic map of cities, lakes, reservoirs, oceans, and land surfaces across the planet and to also map the ocean seabed to a depth of 30m (100ft) in coastal waters and deep estuaries. A laser system on the satellite accurately measures the changing elevation of ice sheets and flows allowing detailed maps to be compiled showing how these are shifting over time. A pulse-mode on the laser also provides accurate ice melt sheet height as it changes over time.

Sea level change

Measurement of sea-level rise was the task of TOPEX/Poseidon, which had been launched on August 10, 1992, a 2,400kg (5,300lb) satellite developed jointly by NASA and CNES, Frances's national space agency. Such was its success that a successor, Jason-1 was launched on December 7, 2001. Designed specifically to measure sea-level change, it carried an altimeter to track the expanding hotter parts of the world's seas and oceans, sea-level height rising and falling across different parts of the globe as heat plumes in the ocean migrate across the surface.

By the time it reached the end of its life in 2012, Jason-1 had been replaced by OSTM/Jason-2, third in a continuing series of satellites measuring ocean topography. A joint venture between NASA. NOAA, CNES and EUMETSAT, the 510kg (1,120lb) satellite was launched on June 20, 2008 and continued to operate for 11 years. In a circular orbit of 1,336km (830 miles) inclined 66° to the equator, it would cover 95 per cent of the Earth's ocean every 10 days, co-ordinating measurements from Jason-1 while that satellite was still operating.

The same partnership continued this vital work with the launch of Jason-3 on January 17, 2016, a 553kg (1,219lb) satellite placed in a similar orbit to its predecessor. Precise measurements of sea level demand a highly accurate measurement of the satellite's altitude and that is obtained via a modified GPS network (see Chapter 12) so that it can know its own position to within 1cm (0.4in). The measurement of sea level is logged in millimetres, but this is averaged over several orbits so that the difference in the satellite's height is evened out.

Findings from the TOPEX/Poseidon to the Jason 3 results show that sea levels have risen by 9.2cm (3.7in) since 1993. Stimulated by this work, further research from oceanographers indicates that seas have risen by 25.4cm (10in) since 1880 but that the rate at which it is increasing has remained constant over the last 140 years. If this is true and the rate remains so, by the end of this century sea levels will be 24cm (9.4in) higher than they are today.

But this is likely to be an extremely optimistic level and discounts the effect as increased meltwater from depleted glaciers and ice shelves adds volume, which could double that figure as the Earth becomes hotter and the melt rate increases. This phenomenon reduces the amount of snow and ice across the Earth's surface and as less sunlight is reflected back into space, more energy is retained in the atmosphere, which adds to a warming planet. The two effects of a rapidly warming atmosphere are increased volume of the seas and oceans as the water expands, mixed with a greater amount of water added as the ice and the glaciers melt away.

It is a continuing study and one joined by Jason 3, which is also equipped with reflectors so that laser beams fired from Earth can obtain the satellite's position with high accuracy. This technique is used to measure the pace at which

RIGHT • *Launched in 2018, Icesat-2 continues the work of its predecessor in measurement of global ice mass, melt rates and surface areas where glaciers and flows are retreating. (NASA)*

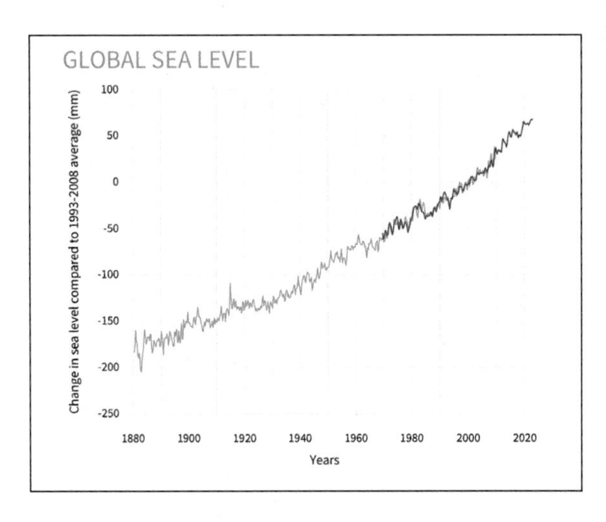

continents are drifting apart, which varies widely but the North American and Eurasian plates are separating at about 2.5cm (1in) a year. Satellite and laser altimetry together with laser beams bounced off the lunar sphere using reflectors left by Apollo astronauts also determines the annual rate at which the Moon is receding from the Earth, about 3.78cm (1.49in), which is the growing rate for human nails.

Some would say late in the day, the European Commission has joined forces with EUMETSAT, ESA, NASA, and NOAA to fund the Sentinel-6 ocean observation programme, a part of Europe's Copernicus environmental satellite initiative, to measure global sea rise. Sentinel-6 is big, with a weight of 1,192kg (2,628lb) and a box-like structure with a length of 5.13m (16.8ft) and a width of 4.17m (13.7ft) by 2.34m (7.7ft), with solar cell arrays providing 891W of electrical power. Launched on November 21, 2020, Sentinel-6 occupies a low-Earth orbit at an altitude of 1,336km (830 miles) and an orbital inclination of 66°. It provides ocean observation from essentially the same path as the Jason satellites and has been fully operational since 2021.

Pioneers together

This shift towards a more focused environmental assessment was pioneered by Japan's Greenhouse Gases Observing Satellite (GOSAT), also known as Ibuki, which means 'breathing' or 'presence' and was configured to observe in the near infra-red and thermal infra-red region of the spectrum. It also has a spectrometer measuring different greenhouse gas concentrations at a very high resolution. The emergence of GOSAT marked a seminal point in Earth observation, where concern about the hydrocarbon emissions played a central role in determining the satellite's purpose.

Launched by Japan's H-IIA rocket on January 23, 2009, the satellite had a weight of 1,750kg (3,860lb) and consisted of an elongated box-like structure, three-axis attitude stabilised in space and with two opposing solar cell wings providing 3.8kW of electrical power. Data was acquired every three days from its sun-synchronous orbit at an altitude of 675km (420 miles) and distributed globally for all nations to use in their own studies of hydrocarbons affecting the atmosphere. The mission was a great success and was followed by GOSAT-2 on October 29, 2018.

Also launched in 2009 was NASA's Orbiting Carbon Observatory (OCO), which was sent up February 24 but lost when the payload shroud failed to separate and, lacking the speed necessary, fell back to destruction in the atmosphere. Its successor OCO-2 was launched successfully on July 2, 2014, to a 98.2-degree sun-synchronous orbit and an altitude of 702km (436 miles). OCO-2 had a length of 2.12m (6.9ft), a width of 0.94m (3ft) and a weight at launch of 454kg (900lb), with solar panels providing 815W. The science instrument was a three-channel grating spectrometer.

OCO-2 was considered an experimental platform for future satellites designed to continuously monitor carbon dioxide levels as they vary in density around the globe by sampling Earth's land and oceans to build maps of distribution and total levels. The primary function will be to measure how carbon dioxide and molecular oxygen absorb sunlight reflected off the Earth's surface and its three spectrometers measure different colour-absorption levels from which concentrations of the hydrocarbon gas can be calculated.

From this, the dynamic patterns of how the ocean exchanges carbon with the atmosphere can be observed, how seasonal changes vary concentrations of the gas, how fossil fuel plumes migrate across North America, Europe and Asia and how different weather fronts affect distribution. Storms and hurricanes exchange carbon at different quantities and at different rates, while different hemispheres store and mix carbon dioxide in different amounts. OCO-2 measures all these parameters and when integrated with the sun-synchronous and geostationary weather satellites produces a comprehensive view of the world's environment.

When launched, OCO-2 was aligned to lead a trail of five other weather, space science and atmospheric satellites, so that all six could measure the same strip of Earth's surface and integrate data from instruments on those other platforms. Currently, two have been retired from the trail but the remaining constellation of four satellites

ABOVE • *The Aqua spacecraft launched in 2002 continues to provide detailed information on the world's oceans, measuring among other things the amount of dust settling on the surface which changes the reflectance level and increases the temperature. (NASA)*

ABOVE LEFT • *Ice melt and expansion in the volume of water in the world's oceans and seas through increased average temperatures is routinely monitored from satellites as they provide an unblinking watch on deteriorating levels. (NOAA)*

BELOW • *Satellites play a vital role in monitoring disasters such as the Deepwater Horizon oil spill of 2010 which seriously affected the condition of the water and marine life in the Gulf of Mexico. (NOAA)*

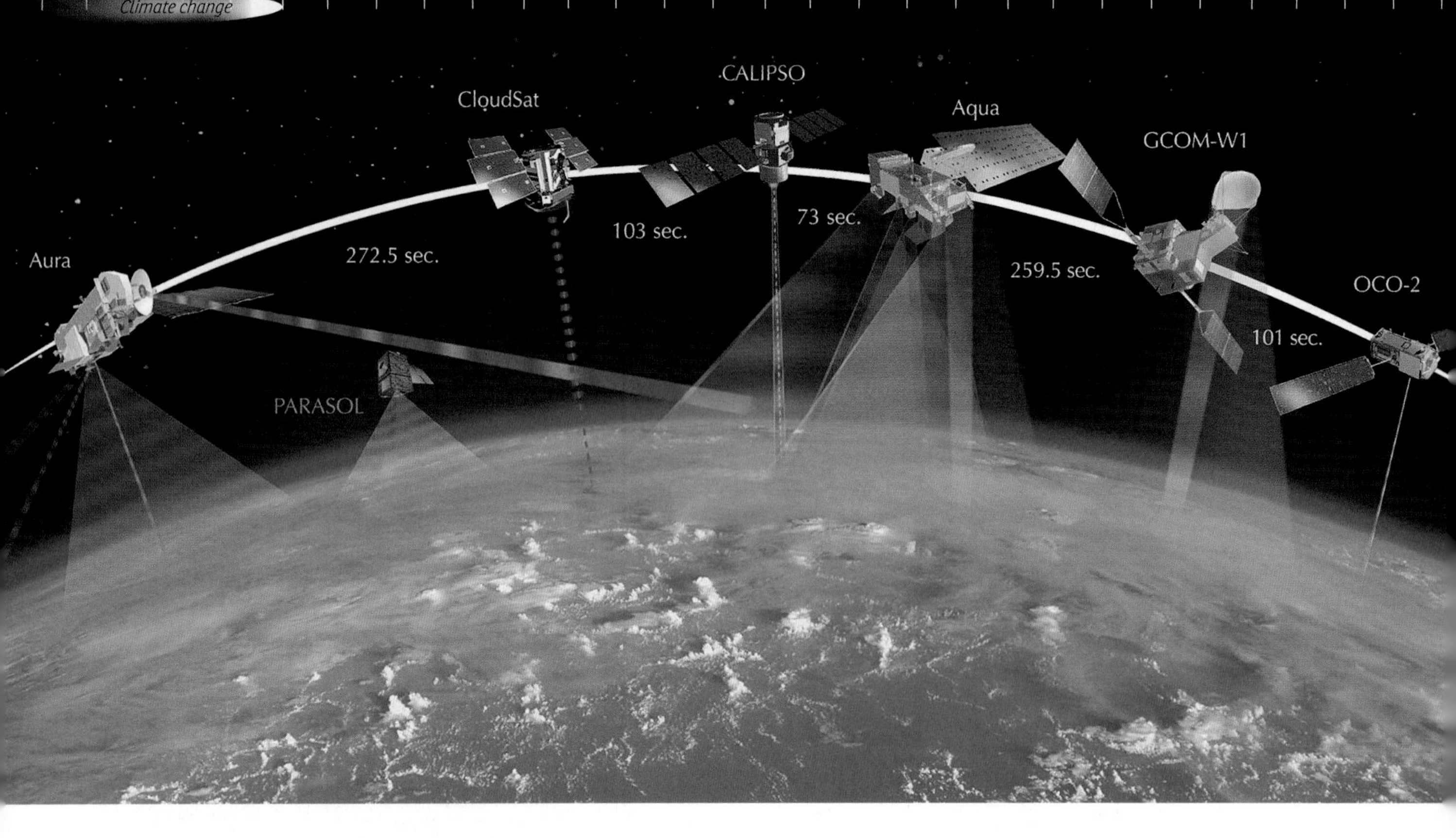

ABOVE AND BELOW • Environmental and climate research satellites are moved around with respect to each other as data-gathering requirements change and new instruments are launched. Spot the different relative positions of these satellites with the changes (below). (NOAA)

continue to provide information across a very wide range of measurements. Spaced just a few minutes apart, they can provide data as though from a single platform. This information is integrated with Japan's GOSAT-2 in a unique venture providing vital data.

European interest

At a government level, European interest in weather satellites, remote sensing from space and the use of orbital platforms for environmental monitoring goes back more than 30 years to the flights of ERS-1 launched on July 17, 1991. With a weight of 2,157kg (4,756lb), ERS-1 was a large satellite built on the design of the SPOT-1 series of commercial remote-sensing satellites and carried a wide range of instruments for measuring sea surface temperatures, cloud temperatures and wind speed and direction. It also carried a large synthetic aperture radar for area surveys of weather over oceans, polar and coastal regions day or night. It was followed by the 2,516kg (5,547lb) ERS-2 on April 21, 1995, with a similar suite of instruments.

Launched on March 1, 2002, Europe's Envisat was the largest environmental Earth satellite launched to date and carried nine science instruments for measuring water vapour, sea surface temperature, the pressure and temperature of the atmosphere, the amount of solar reflection from the surface and the quantity of sea ice. With a weight of 8,211kg (18,102lb) it measured 26m x 10m x 5m (85ft x 33ft x 16ft) and occupied a sun-synchronous orbit at 773km (480 miles). Envisat was

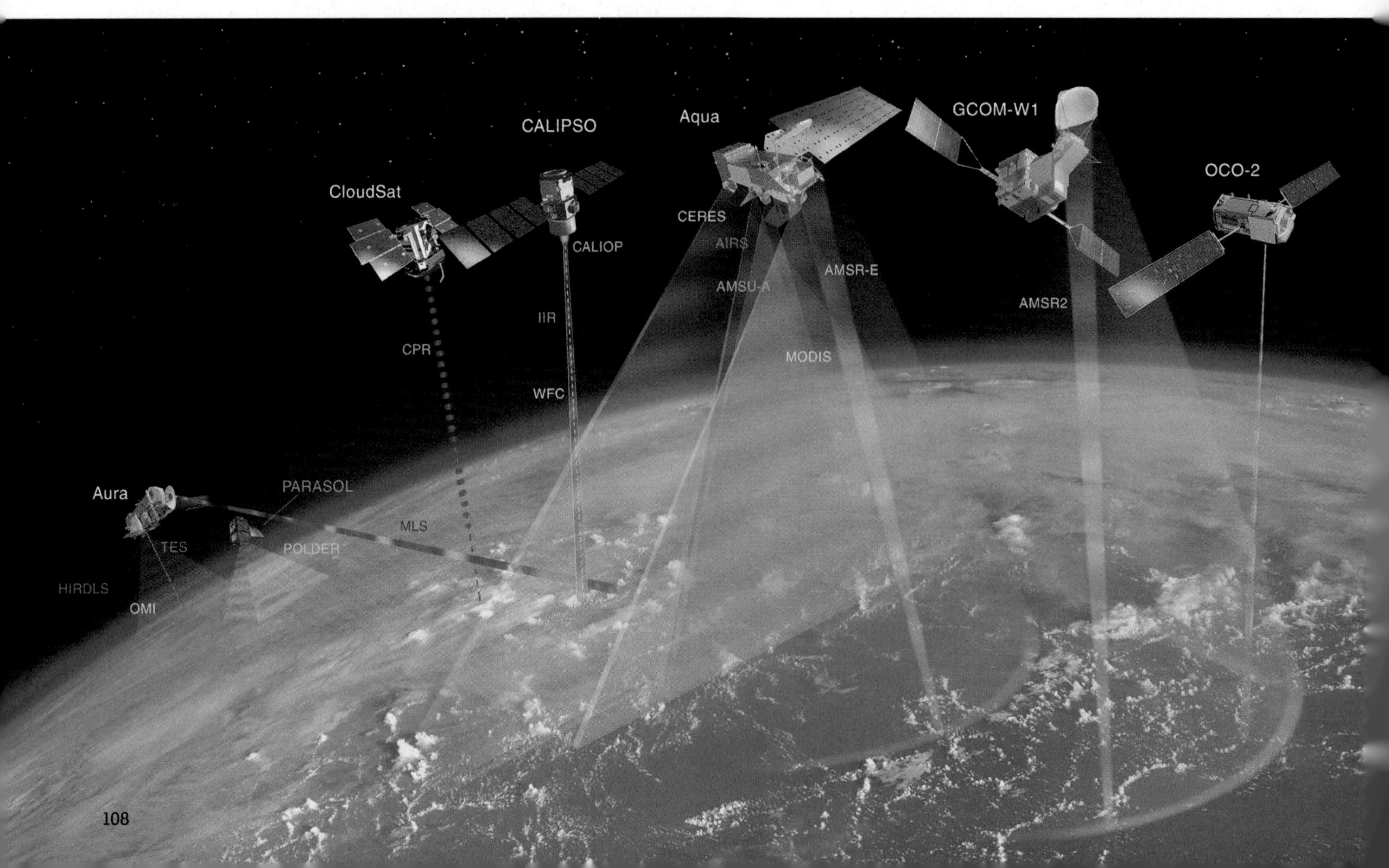

a great success, operated until April 2012 and served as a precursor to the Copernicus programme.

By the late 1990s there was a gathering consensus at ESA and in the European Commission for a major initiative to study the Earth's environment and in 2012 the name Copernicus was chosen for an initial series of seven satellites. In addition to the 27 European member states, Norway, Switzerland, and Iceland agreed to provide their localised environmental data in exchange for information from the Copernicus satellites, each one bearing the name Sentinel, and each configured to provide extensive and complementary data-sets about the planet and the environment.

Launched on April 3, 2014, Sentinel-1A provided radar imaging of land and sea surfaces continuously day and night, and was followed by Sentinel-1B on April 25, 2016, the mission for which ended in August 2022. The two Sentinel 2 satellites were launched on June 23, 2015 and March 7, 2017, configured for high resolution optical imaging of vegetation, soil and water conditions and inland and coastal waterways. Equipped for observing general ocean and land conditions, two Sentinel-3 satellites were sent into space on February 16, 2016 and April 25, 2018. Several Sentinel satellites were designed to provide concurrent data for integrating their observations and more are planned for the years ahead.

Meanwhile, commercial SPOT satellites continue to produce high-resolution data. Launched in 1998 SPOT-4 provided images in four spectral bands, three in multi-spectral mode (green, red, and near-infra-red) and a fourth was in a short-wave infra-red band. The single mono-spectral imager was equivalent to the panchromatic mode on the first three satellites. SPOT-5 went into orbit in 2002 and provides customers with images offering a resolution of up to 2.5m (8.2ft) while SPOT-6, launched in 2012, went up on India's PSLV rocket and provides panchromatic and colour images with a 1.5m (4.9ft) resolution. Its twin, SPOT-7, was launched by PSLV in 2014 and sold to Azerbaijan's space agency which renamed it Azersky.

Monitoring of the Earth's environment is a legacy of spy satellites, resource satellites and weather satellites. Together, all these different platforms have transformed the way humankind views its planetary home, a collective knowledge that vindicates concerns raised in the 1970s about the fragile nature of Earth's ecosystems and its limited resources. Global awareness benefitted from an unlikely source – satellite communication systems that established a unified worldwide network for managing international telephone, telegraph, and television. Like weather satellites before them, and environmental satellites today, communication satellites would also find a global home.

ABOVE • *Jason-2 prior to its launch in 2008 to support US-European studies into sea level changes. (NASA)*

LEFT • *Europe's Envisat captures a rare view of the UK without significant cloud cover. Images such as this provide vital data for urban and rural planning and monitoring of the built environment. (ESA)*

TALKING TO THE WORLD

Satellites are key to a technological and cultural transformation of our planet.

BELOW • Launched on December 18, 1958, Project Score consisted of an Atlas B equipped with a communication package to send to Earth a message from President Eisenhower after it had been placed in orbit. (USAF)

BELOW RIGHT • The Echo II satellite was the second of two NASA-funded passive relay satellites in which it served to link two places without re-amplifying the signal. (NASA)

The telephone has been one of the most empowering tools of the modern world. Either by landline or wireless, communication over great distances has revolutionised the way people live, work, and connect. It began in 1876 when Alexander Graham Bell picked up a microphone connected to a wire and uttered the first words by telephone: "Mr Watson, come here I want to see you." We do not have an accurate record of the reply but that doesn't matter. The message was enough to change the world, albeit very slowly at first as initially this novelty was exclusively the preserve of the wealthy.

While landlines were the standard way of connecting people, wireless communication first occurred on February 19, 1880, with the invention of the photophone, later known as the radiophone, where voices were conveyed on beams of electromagnetic radiation. It soon became standardised as radiotelephony or RT. That opened the practicality of radiotelegraphy, the transmission of telegrams by radio, and so was completed the cycle of adapting communication from a line to a wave.

Yet it was British science fiction writer Arthur C Clarke who, in the February 1945 edition of *Wireless World*. had a letter published proposing that a satellite placed over the equator at a distance of 36,000km (22,400 miles) would appear to remain fixed in space and provide a perfect relay platform for communications. Three such geostationary satellites placed at 120-degree intervals around the globe would provide total coverage for every country on Earth. Clarke imagined that this utopian possibility would take a further 50 years to achieve when, in fact, it would take less than 20.

Clarke circulated a paper three months later, in which he expanded on the technical challenges and the possibilities for such satellites. This was reprinted in *Spaceflight*, the monthly magazine of the British Interplanetary Society in its March 1968 issue and a copy of that can be obtained from them. In it, Clarke recalls how he formulated the concept drawing on his experience in charge of ground control approach (GCA) radar during the war. But Clarke imagined that a telephone relay station in space would require personnel to operate it.

Clarke had not been the first to suggest a geostationary orbit, but his growing reputation as a futurist thinker helped project the concept. In 1954, AT&Ts John R Pierce delivered a talk in which he evaluated the technical and economic consequences of such a programme. This was followed by an explanatory article the following year. Pierce considered a series of stages, beginning with a mirror from which signals could be bounced for collection by another radio station in line-of-sight. Following would be a repeater satellite in a medium-altitude orbit and capable of boosting the signal before re-transmitting it to another station on Earth. Only when that technology had been mastered did Pierce imagine a geostationary satellite as prophesied by Clarke.

Slow evolution

But communications via satellite would evolve slowly and progress along a slightly different track to that imagined. After Sputnik I shocked the world in October 1957, the race was on to express US technological ability to counter anything the Soviets could achieve. When in 1958 President Eisenhower established the Advanced Research Projects Agency (ARPA), tasking it with taking charge of all the diverse, non-classified space projects then underway or conceived, he provided a propaganda opportunity that saw the first demonstration of space-to-Earth voice communication.

On December 18, 1958, an Atlas-B missile equipped with transmission equipment for broadcasting a pre-recorded message to Earth was launched from Cape Canaveral. Depleted of propellant, the entire missile was placed in orbit, from where it transmitted a message to listeners. With no payload other than this transmission equipment and a tape recorder, it was a low-key attempt at boosting the profile of the United States and had no technical value.

Nevertheless, in the period before Christmas it was a very American message: "This is the President of the United States speaking. Through the marvels of scientific advance, my voice is coming to you from a satellite circling in outer space. My message is a simple one: Through this unique means I convey to you and to all mankind, America's wish for peace on Earth and goodwill toward men everywhere." In fact, the original message was from a US Air Force officer and Eisenhower intervened shortly before the launch when the recorder had already been loaded. In orbit it was reloaded electronically with the president's voice.

Satellites for communication by voice, picture or data can be useful in several different orbits. Very low orbits will carry the satellite from horizon to horizon quickly, allowing little time for a ground station to lock on before it disappears again. A medium-Earth orbit slows the transit time to a fixed observer, while a high-altitude orbit delays the time even further. Only a geostationary orbit provides a fixed point in space for sustained communications between two distant sites on Earth. However, the further out the satellite, the less is the strength of the signal when it reaches the ground. Radio waves are subject to the inverse square law, where the strength of the signal is inversely proportional to the square of the distance.

Some compensation can be achieved through what is known as beam-shaping, where the signal radiated from the satellite is focused more narrowly rather than being sent out in all directions. In this way, the same amount of radiated energy is constrained to a smaller area on the surface where the beam is focused. This also prevents different frequencies interfering with each other.

The net result is a spectrum of capabilities that for several years in the early phase of the Space Age relied heavily on the ability of rockets and satellite launchers to achieve the desired orbits. As noted earlier, Clarke had imagined that his geostationary relay satellites would be manned, but getting to that orbit proved a challenge for both rocket and the platform it launched. By the time it was achieved the degree of progress in electronics and systems that could be remotely controlled, pre-programmed, or autonomous in their operation made the expensive astronaut redundant for those tasks.

Passive and active

There are two types of communication satellite, passive in which the signal is reflected back to the ground and active, in which the signal is boosted before being retransmitted to a station on the surface. Passive signals are only a fraction of the strength by the time they get reflected back, while boosted signals are much stronger, require smaller ground antennas and can carry more information.

Work on the development of passive satellites began in the early 1950s with the US Navy, always at the forefront of radio engineering and development, vital for maritime communications and in the US naval tradition

ABOVE • *Launched on July 10, to an elliptical orbit, the Telstar satellite provided tests with an active relay system and transmitted the first TV pictures across the Atlantic Ocean. (Author's collection)*

BELOW • *The horn antenna for Echo satellite relay located at Holmdel, New Jersey. (NASA)*

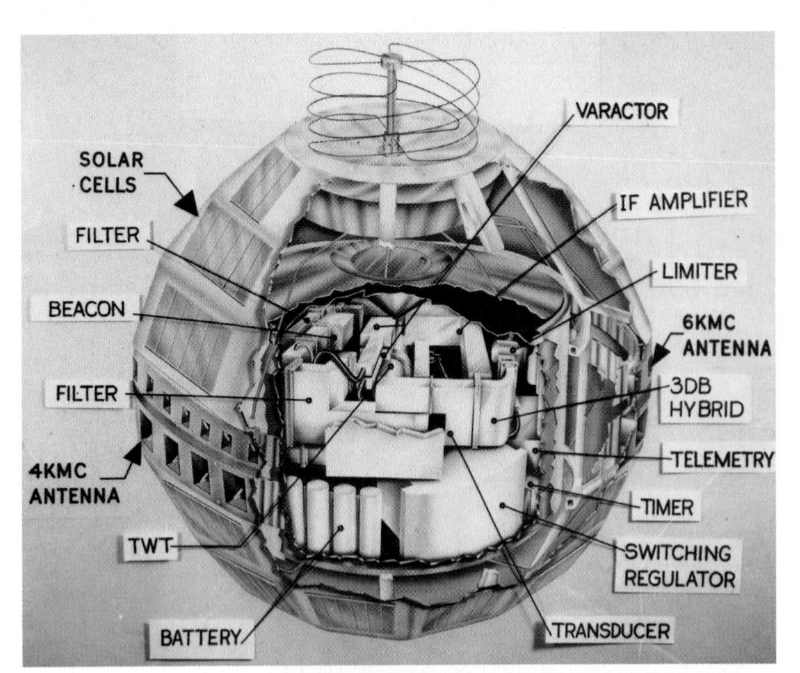

ABOVE • The interior layout of the Telstar experimental communications and TV satellite. (Bell Labs)

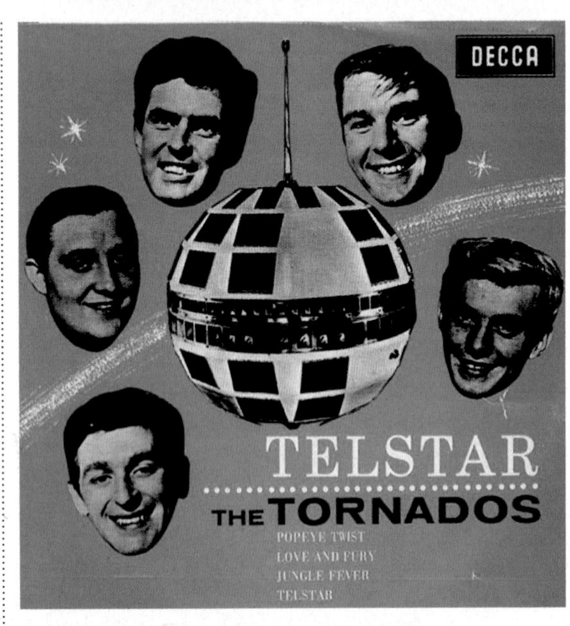

RIGHT • Written for the successful flight of Telstar, the British pop group The Tornados got a had success with this instrumental using the Jennings Univox. (Decca)

of conducting scientific research. In 1951, the Navy had achieved a first for telecommunications by bouncing a signal off the Moon and during that decade it made some of the more significant steps towards a working communications satellite concept.

With the acceleration of the space programme and the formation of NASA in October 1958, it fell to the civilian space agency to launch the world's first passive communication satellite, the ball-shaped Echo I on August 12, 1960. Sent to an orbit of 1,524km (947 miles) by 1,684km (1,046 miles) inclined 47.2°, it was used to bounce signals between stations at Goldstone Dry Lake, Mojave, California and Holmdel, New Jersey.

With an inflated diameter of 30m (98ft), it had a skin only 12.7μm (0.00050in) thick. Echo I weighed 71kg (157lb) at launch and was inflated by a small quantity of gas. It proved what everyone knew it would – that it was possible to bounce radio waves off a spherical structure in space. But it performed valuable service in providing data on the atmospheric density and solar pressure at extreme altitude. It was followed by the much larger Echo II with an inflated diameter of 41.1m (135ft) launched on January 25, 1964. This had a rigidised skin of 9μm (0.00035in) Mylar sandwiched between two layers of 4.5μm (0.00018in) aluminium foil.

While a fascinating exercise in reflecting radio waves off a balloon, Echo would never be the forerunner of true satellite communications but the ability to launch into, and operate from a geostationary orbit, was still some way off. But less than two months after the launch of Echo I, on October 4, 1960, the US Air Force launched Courier 1B to an elliptical orbit of 1,237km (769 miles) by 938km (583 miles) with an inclination of 28.33°.

Developed by the US Army Signal Corps, Courier 1B was a 129cm (51in) diameter spherical satellite weighing 225kg (496lb) with four transmitters and five tape recorders, of which four were digital. It worked for 17 days, demonstrating UHF communications, and relaying a message from President Eisenhower to the United Nations.

Around this time in the United States, Bell Telephone Laboratories and NASA formed an agreement with the General Post Office in the UK and with the National PTT in France to jointly support an experimental TV satellite. Bell would build the satellite, to be named Telstar, NASA would launch it and the BBC in the UK would co-ordinate operations from the station at Goonhilly Downs, Cornwall. It would operate with TV conversion equipment developed at Television Centre in London. From the Andover Earth Station in Maine, signals would be sent to Cornwall and from there via the satellite to stations in Germany and Italy.

Launched on July 10, 1962, Telstar 1 transmitted the first commercial TV image by satellite between Andover and Pleumeur-Bodu in France. On July 23, TV pictures were sent to Europe and broadcast live on Eurovision networks across the continent and the United States, national broadcasters showing its first images of the Statue of Liberty and the Eiffel Tower. In an unexpected use of satellite TV to quell increasing alarm over rumours of an imminent devaluation of the dollar, casual remarks by President Kennedy to the contrary reinforced confidence and stabilised world markets. A global first for satellites!

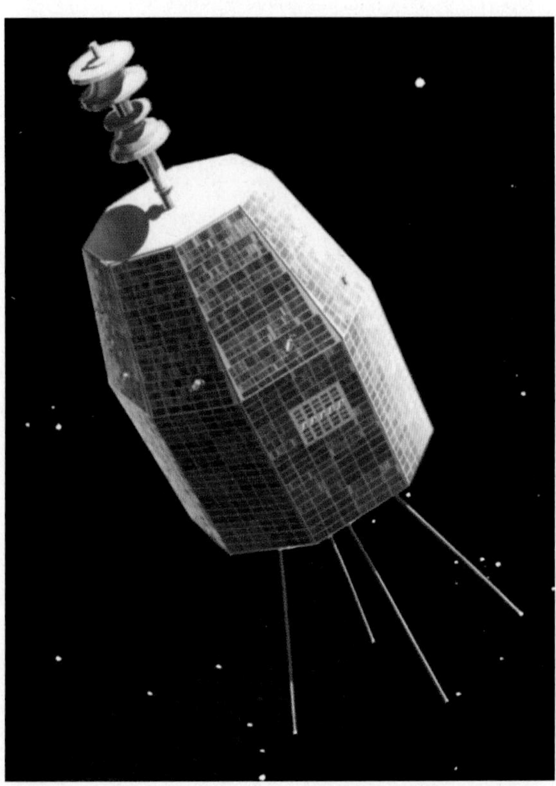

RIGHT • Built for NASA by RCA, Relay-1 was launched in December 1962 and successfully transmitted the first TV programme between Japan and the continental US. (NASA)

FAR LEFT • *The 'father' of the geostationary, spin-stabilised satellite, Harold Rosen was the driving force behind a successful string of Hughes communication satellites which would trigger an expanding commercial business. (HAC)*

LEFT • *The Hughes Syncom satellites were technology demonstration platforms with the second, shown here, being the first to achieve geosynchronous orbit in July 1963. (HAC)*

Further tests were made by using the satellite to synchronise clocks on both sides of the Atlantic, relaying data to an IBM 1401 computer in France and demonstrating voice communication. Although small, with a diameter of 88cm (34.5in) and weighing only 77kg (170lb) and solar cells delivering only 14W or electrical power, Telstar had several novel features and a lot of capability. With a new travelling wave tube amplifier (TWTA), it could handle a TV signal, telephone circuits or a single block of data. Unfortunately, the satellite succumbed to the effects of nuclear tests in space, its transistors being fried by radiation, which only served to emphasise the need for military satellites to be hardened as defence against these effects.

Higher and higher

The effect of the satellite on the world was dramatic, with news coverage unprecedented for its time in a year when NASA was launching astronauts into space, sending probes to Venus, planning flights to Mars and Russia was starting to put missiles in Cuba. Culture was changing and a new generation was aligning the 'swing' age with the greatest technical achievements of the day. The British instrumental group Tornados made a chart-topping record, simply called Telstar with a futuristic electronic sound created by a Jennings Univox. It carried the mood of the age with five million copies sold and could be heard on juke boxes around the world.

Identical to its predecessor, Telstar II was launched on May 7, 1963, by which time another experimental satellite had been launched. Telstar had operated from a highly elliptical path of 5,933km (3,687 miles) by 952km (592 miles) at an inclination of 45° to the equator, a complete orbit taking 2hr 37min. NASA wanted its own research and development satellite and contracted RCA to produce Relay, which had been launched on December 13, 1962 to an orbit of 7,438km (4,622 miles) by 1,322km (821 miles) and at an inclination of 47.48°.

The shape of the 170kg (375lb) Relay satellite was designed to fit within the launch vehicle's payload shroud. It consisted of an eight-sided prism with a maximum diameter at the base of 73.7cm (29in) and a height of 48cm (19in), on top of which was a mast-like structure with a length of 46cm (18in), carrying the antenna systems for microwave communications. Power was provided by solar cells generating 40W, but a 250W-hr battery made up the difference to reach a requirement for 120W.

Blighted by technical problems, Relay 1 was nevertheless the first satellite to broadcast live TV between the continental US and Japan and in 1964 was used to transmit the summer Olympics from Tokyo. It did much to confirm expectations regarding the development of satellite-based telecommunications and stimulated a great deal of progress with ground stations, co-ordination of international services and the technology for satellites of this type. With a weight of 184kg (406lb), Relay 2 was launched on January 21, 1964, by which time the first geostationary satellite had already been launched.

Patently, the technology for using satellites to relay signals from one side of the Earth to the other was within reach. Neither was there anything new about the concept of a satellite orbiting at a distance which would take it 24 hours to go once around the planet and therefore appear stationary. But nobody had done it. The idea of a geosynchronous

BELOW • *The first Intelsat satellite named Early Bird went into orbit in April 1965, first in a continuing line of telecommunication satellites encircling the globe with international connections among almost all the nations of the world. (Intelsat)*

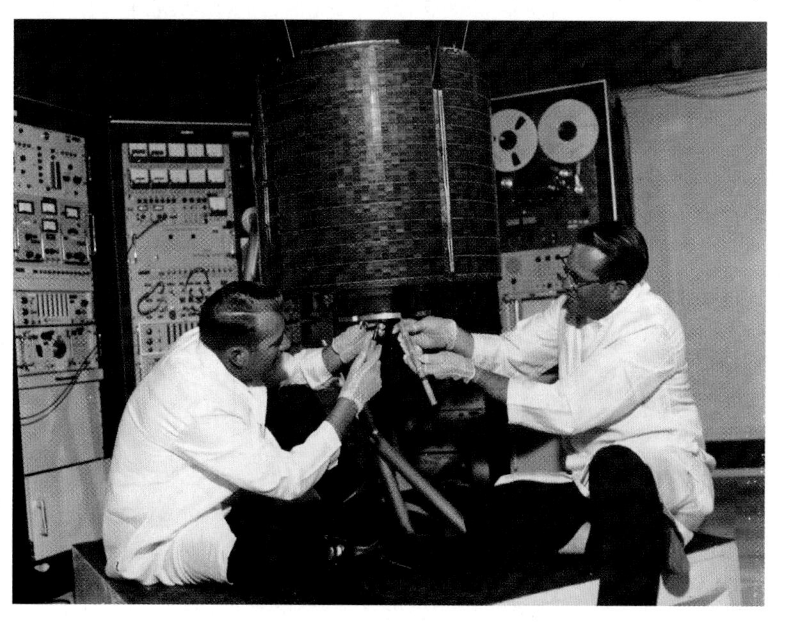

ABOVE • *Launched to support Pacific Ocean communications traffic, the second in the Intelsat 2 series was named Lani Bird after the extinct species of Hawaiian honeycreeper. (Intelsat)*

satellite is first recorded in the only book by Austro-Hungarian army officer Herman Potočnik, published in 1928 as a discourse on space stations and interplanetary travel. But it was Arthur C Clarke who popularised the concept in 1945. In more recent times, the geostationary orbit is named after this visionary – the Clarke orbit.

The terms geosynchronous and geostationary are frequently interchanged, but there is a significant difference between the two. Geosynchronous satellites are those in which the time they take to go around the planet is the same as the Earth spins in one revolution but with the inclination of the orbit at a tangent to the equator. Because of this, over a 24-hour period, the satellite appears to drift north or south depending upon the orbital inclination, making large loops above a fixed spot, or tracing out a figure-of-eight path. Geostationary orbits are where the orbital inclination is on precisely the same plane as the equator, from where they appear to remain fixed.

Subtly difference

The subtle operational difference between the two is that for geosynchronous orbits, ground antennas must track out the apparent movement of the satellite, whereas they need to move very little for geostationary satellites – hence the term. The obvious advantages of the latter were recognised, and the development of geostationary satellites became a prime mover

RIGHT • *Launched on March 11, 1990, by Titan rocket the defunct upper stage left Intelsat VI stranded in low-Earth orbit. A Shuttle mission in 1992 captured the satellite, added a suitable boost motor, and sent it to geostationary orbit. (NASA)*

in advancing the capability for global coverage. Three orbital locations at 120° around the planet would cover the whole Earth, apart from extreme north and south latitudes, where the curvature of the horizon is so great as to place the observed position of the satellite as being too low on the horizon.

While Arthur C Clark is popularised as the moving influence in the development of geostationary satellites, the first steps to make it possible came from a completely different source. A brilliant engineer working with electronics company Raytheon, Harold Rosen, had never heard of Clarke but he read a paper by two engineers from Bell Labs, John Pierce, and Rudy Kompfner. In 1959, they published a review and judged the technology for a geostationary satellite too risky, citing the lack of launch vehicles to achieve that orbit, proclaiming that they would be far too heavy to launch and that their inherent complexity would prevent them from becoming a commercial success.

It was the classic case of professional hubris: if engineers are unable to get to their machines to tinker, they will fail because they cannot possibly be made to operate on their own for several years unattended. Incensed by this, Rosen, who had now moved to Hughes, believed that with its business wholly involved with ground facilities, Bell Labs had no concept of dramatic weight reduction and miniaturisation to make them possible. Moreover, he would not accept the logic of unattended failure, proclaiming that if engineered correctly, satellites would achieve greater reliability simply because engineers could not get to them!

Despite a wall of opposition to the idea, Rosen took the idea to his former employer but rather than lose him to Raytheon, he was allowed to start a development project with some innovative ideas. And so, Hughes got a head start in a radical new business which would quickly place them centre-stage. To keep weight as low as possible, attitude control would be through spin-stabilisation and miniaturised, low-mass electronics and structural systems would support a design, known as Syncom, to be within the payload capabilities of a Delta launch vehicle.

ABOVE • Successive generations of increasing efficiency and capability evolved through to the Intelsat IV/IVA series launched between 1971 and 1978. (HAC)

LEFT • Maritime telecommunications support began with the European Marecs series of satellites to serve a market in desperate need of satellites to connect ocean-going fleets with shore installations, updating electronic maps and guiding ships through adverse weather. (ESA)

The advantages of a geosynchronous satellite were obvious, but risk was high due to the pioneering nature of the concept and NASA engaged in a support programme to assist Hughes with its development. Telstar and Relay satellites were effective but too limited in capability and in their medium-Earth orbits were unsuited to a commercial programme. Considerable experience had been gained with these existing satellites and that fed across into the semi-commercial programme, with a view to encouraging traffic from telecommunications agencies and even broadcasting companies. But they were demonstrators, with capacity for one, two-way telephone connection or 16 two-way teletype links.

Three satellites were planned in the HS-301 series, each drum-shaped with a diameter of 71cm (28in) and a height of 39cm (15in). When measured from the base of a small orbit-insertion motor it had a total height of 64cm (25in) to the top of its slotted-array antenna. Weighing 68kg (150lb) at launch, it carried rocket propellant for a small motor to circularise its orbit, after which the satellite would weigh 35.5kg (78lb).

Syncom I was launched on February 13, 1963, and the Delta rocket placed it on an elliptical path in which five hours later, the small apogee-kick motor designed to circularise the orbit fired for 20 seconds. But all contact was lost. Its twin, Syncom II, was launched on July 26 1963 and placed in the same elliptical path inclined 33° to the equator as its precursor. This time, the kick motor fired and placed it in geosynchronous orbit, the first time that had been achieved.

ABOVE • *Marecs led to Inmarsat satellites with land-based portable terminals and a wide application to both civil and military requirements. (USN)*

Initially, Syncom II was placed in a super-geosynchronous path, which meant that it was slightly out of synch with the rotation of the Earth and that allowed it to drift west at a rate of 4.5° of longitude each day, engineers eventually stopping it over 55° W by commanding its on-board propulsion system to lower the orbit again. NASA and Hughes saw the tests with Syncom II as vital for engaging the public in a demonstration of how effective satellite communication could be and set up 110 events to show that. Symposia, open days at air bases, NASA centres and companies all carried replicas of Syncom and displayed its capabilities.

Launched on August 19, 1964, Syncom III became the first satellite placed in a geostationary orbit, the programmed Delta rocket steering the trajectory to match the orbit with the equatorial, 0° latitude. Much the same as its predecessors, Syncom III had a TV channel, but over the next two years the two surviving satellites did a lot of tests for the military, which took over responsibility for the series on January 1, 1965. It helped prepare them for the first defence communication satellites, about which more in the next chapter. But the work launched Hughes into an entirely new industry.

The government intervenes

Nature abhors a vacuum and so do politicians. Before the technical and engineering challenges of communication satellites had been demonstrated, on August 31, 1962, President Kennedy signed off the Communications and Satellite Act. Open handed when it came to defining its purpose and responsibilities, it was an attempt to regulate the emerging industry. To serve as a publicly funded operation, the Communications Satellite Corporation (Comsat) began trading on February 1, 1963.

When Telstar, Relay and Syncom demonstrated a potential market for such services, the International Government Organization (IGO) was established on August 20, 1964, with seven countries signed up as members. The 1962 act also established the International Telecommunications Satellite Organization (ITSO), soon known as the name by which it is known today, Intelsat, but within a very different framework. Hughes was clearly a front-runner for building its satellites and the international model was set up around a fair and equitable base.

Although the United States was running the operation, it was only going to work if other countries got behind the agreement and that defined the standards and funding profile within which it was expected to operate. It required countries to sign up through their respective posts and telecommunications agencies. From that point, each country could decide to share the broadcast services and telephone lines with other broadcasters internally but not to use the signals to generate markets outside their own territorial boundaries.

Intelsat expected satellite traffic to grow commensurate with the demand and as possible through the evolving technology. It based its business model on a concept of pay-as-you-buy, which required each signatory to pay a percentage portion of an annual running cost proportional to the number of half circuits it used. A half circuit is a one-way voice channel, a full circuit being a two-way conversation. That way, each country need only pay what it could afford to put in. That set a firm income for procuring the satellites while encouraging growth, since only the amount used had to be paid for.

RIGHT • *Throughout the 1970s and 1980s, several countries purchased US-built satellites for launch on US rockets, the first being Telesat Canada with the Anik 1 sent to geostationary orbit in November 1972. (HAC)*

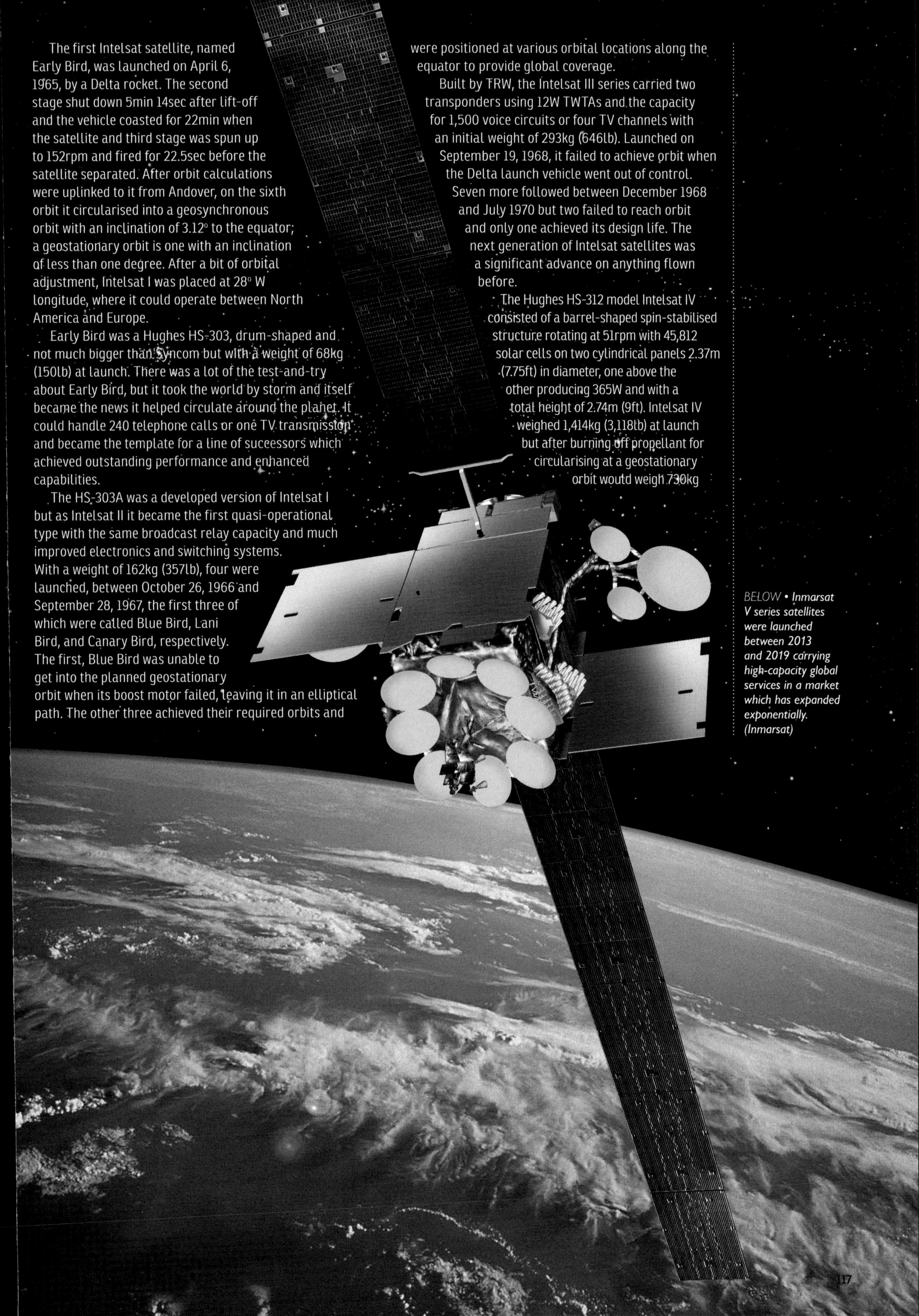

The first Intelsat satellite, named Early Bird, was launched on April 6, 1965, by a Delta rocket. The second stage shut down 5min 14sec after lift-off and the vehicle coasted for 22min when the satellite and third stage was spun up to 152rpm and fired for 22.5sec before the satellite separated. After orbit calculations were uplinked to it from Andover, on the sixth orbit it circularised into a geosynchronous orbit with an inclination of 3.12° to the equator; a geostationary orbit is one with an inclination of less than one degree. After a bit of orbital adjustment, Intelsat I was placed at 28° W longitude, where it could operate between North America and Europe.

Early Bird was a Hughes HS-303, drum-shaped and not much bigger than Syncom but with a weight of 68kg (150lb) at launch. There was a lot of the test-and-try about Early Bird, but it took the world by storm and itself became the news it helped circulate around the planet. It could handle 240 telephone calls or one TV transmission and became the template for a line of successors which achieved outstanding performance and enhanced capabilities.

The HS-303A was a developed version of Intelsat I but as Intelsat II it became the first quasi-operational type with the same broadcast relay capacity and much improved electronics and switching systems. With a weight of 162kg (357lb), four were launched, between October 26, 1966 and September 28, 1967, the first three of which were called Blue Bird, Lani Bird, and Canary Bird, respectively. The first, Blue Bird was unable to get into the planned geostationary orbit when its boost motor failed, leaving it in an elliptical path. The other three achieved their required orbits and were positioned at various orbital locations along the equator to provide global coverage.

Built by TRW, the Intelsat III series carried two transponders using 12W TWTAs and the capacity for 1,500 voice circuits or four TV channels with an initial weight of 293kg (646lb). Launched on September 19, 1968, it failed to achieve orbit when the Delta launch vehicle went out of control. Seven more followed between December 1968 and July 1970 but two failed to reach orbit and only one achieved its design life. The next generation of Intelsat satellites was a significant advance on anything flown before.

The Hughes HS-312 model Intelsat IV consisted of a barrel-shaped spin-stabilised structure rotating at 51rpm with 45,812 solar cells on two cylindrical panels 2.37m (7.75ft) in diameter, one above the other producing 365W and with a total height of 2.74m (9ft). Intelsat IV weighed 1,414kg (3,118lb) at launch but after burning off propellant for circularising at a geostationary orbit would weigh 730kg

BELOW • Inmarsat V series satellites were launched between 2013 and 2019 carrying high-capacity global services in a market which has expanded exponentially. (Inmarsat)

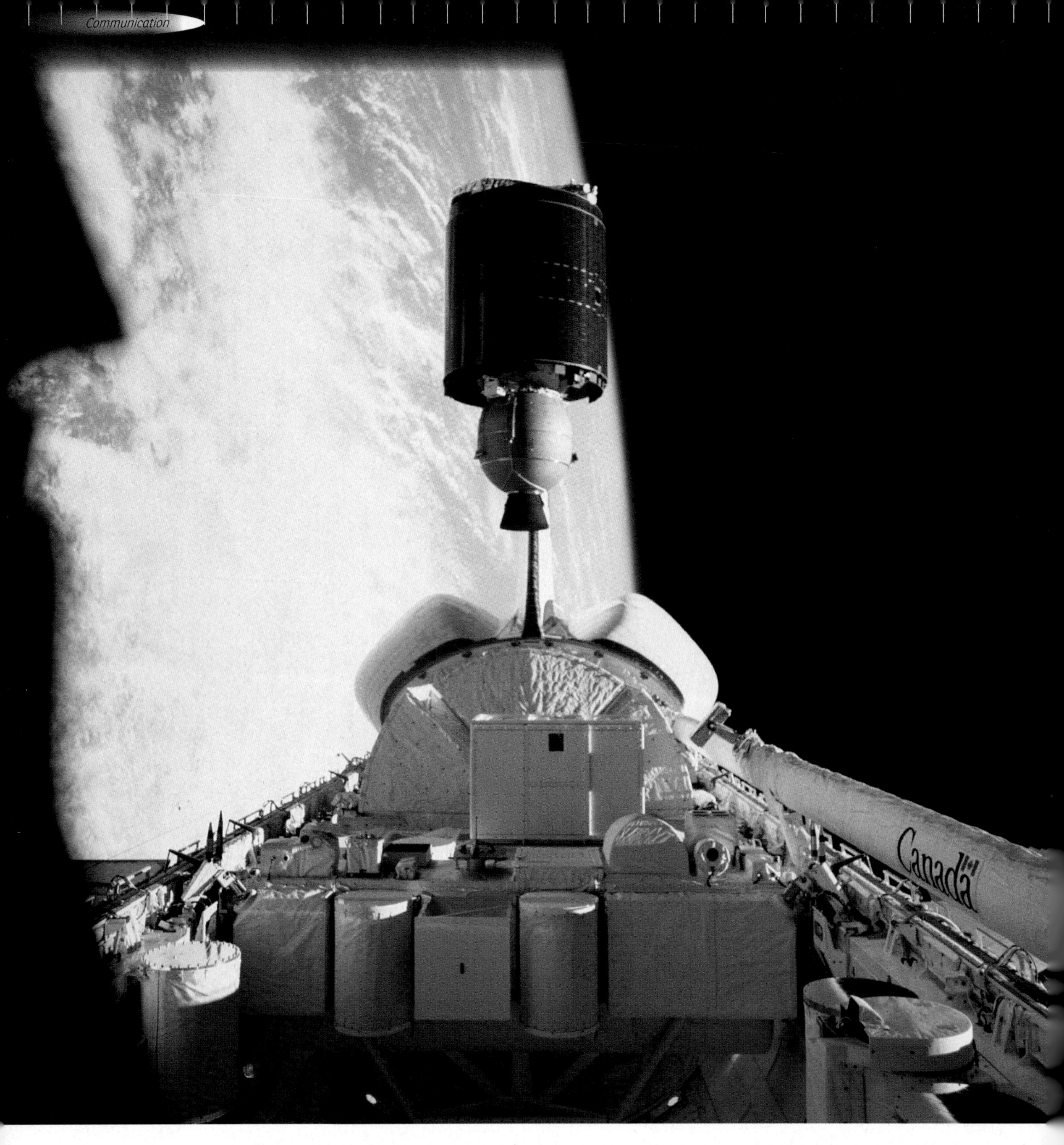

ABOVE • ***Anik C-2 being launched out of the Shuttle in 1983 attached to a boost stage which would propel it to geostationary orbit. (NASA)***

(1,609lb). Each satellite could handle 6,000 voice circuits and up to 12 simultaneous TV channels. Seven of the eight flown between January 1971 and May 1975 were successful.

Spin peculiarities

The peculiarities of a spin-stabilised design require one part to spin in the opposite direction to ensure that the platform that holds the antennas remains pointing in a fixed direction at the Earth. The join between these two contra-rotating sections is known as the bearing and power take-off assembly, or BAPTA, across which the electric power of the spinning section must be transferred to the de-spun section for powering the satellite's systems and communication equipment. This was but one of the technical challenges successfully met by satellite designers.

Further development of the HS-312 produced the more capable Intelsat-IVA which was essentially the same as its predecessor but with almost twice the capacity, by using the same frequencies directed through four separate spot beams, rather than two on Intelsat IV. Instead of two spot beams, one covering North America, the other Europe, there were now spot beams for South America and Africa as well. Other zones around the world would have similar beam separations using the same frequency for different footprints, the area on the Earth's surface covered by the beams. This was the latest evolution that would become standard for all communication satellites.

The last in the series built by Hughes, Intelsat IVA, was essentially the same structural configuration with a diameter of 2.38m (7.75ft) and a total height of 7.01m (22.9ft) when fully deployed. It had a launch weight of 1,515kg (3,335lb) reduced to 825kg (1,819lb) after its orbit motor had done its job. Both Intelsat IV types were launched by Atlas-Centaur. Five of the six IVA types launched between September 1975 and March 1978 were successful, the penultimate flight ending when the upper stage caught fire after launch and before it could achieve orbit.

Due to the four-beam configuration, each Intelsat IVA could handle 12,000 voice-grade channels through 20 transponders compared to 12 for the Intelsat IV, with four wide-area beams and 16 for the spot-beam footprints. The bulk of satellite communications were across the Atlantic Ocean, but the design of the Intelsat IVA took account of a rapidly expanding Indian Ocean and Pacific Ocean traffic. The success of the overall service was exponential, and the IVA provided the capacity sought by an increasing number of Intelsat member states.

The size and capacity of communications satellites grew quickly across the turn of the century, with manufacturers developing standard satellite bus configurations capable of adaptation to customer requirements. Typical is the SSL 1300 bus built in the US by Maxar Technologies. Developed several times for increasing demands, derivatives of the SSL 1300 were capable of producing 10kW of radiated power served through up to 150 transponders with a launch weight of 6,700kg (14,774lb) and a width when folded up and encapsulated in the launch vehicle's payload shroud of up to 5m (16.4ft).

As traffic grew and capacity expanded, the clamour for competition and commercial involvement brought an end to the inter-governmental public ownership and signatories agreed, certainly in their own interest, to move toward privatisation. The US Congress passed a law allowing that and on July 18, 2001, Intelsat entered private ownership. Four years later, it was sold to four private equity firms and in 2013 it was renamed Intelsat S.A. Seven years later, it filed for bankruptcy but emerged in 2022 stronger and with a portfolio of new services and an invested interest in new technology.

All at sea

With direct coverage available for land-based stations and TV connections for terrestrial broadcasters and telecommunication agencies, there was one very big, yawning gap. No coverage for two-thirds of the Earth's surface covered by water. With the UK and Russia, the largest maritime operators in the world at the time, the International Maritime Satellite Organization (INMARSAT) was set up on September 3, 1976, effective from 1979 and with its headquarters in London. Its remit was to provide maritime communication and rescue-support services on the same not-for-profit, inter-governmental basis as Intelsat.

The first satellites were named Marisat, three being launched between February and October 1976 and each a Hughes HS-303 design weighing 362kg (798lb) in geostationary orbit. Spin-stabilised, they provided maritime communications in three discrete bands for ship-to-satellite and connections between the satellite and shore stations. Placed around the globe at 15° W, 176.5° E and 72.5° E, they pioneered the concept and laid the groundwork for future operations.

With 80 per cent of Europe's trade moving by sea, the European Space Agency instituted the Marecs programme, aiming to operate satellites for Inmarsat, which it achieved with two successfully operational in 1981 and 1984 for Atlantic Ocean and Pacific Ocean coverage, respectively. Marecs satellites were launched on ESA's Ariane rocket from French Guiana. A third launched in 1982 failed to reach orbit. Based on a three-axis stabilised communications satellite and built by British Aerospace, they carried 35 two-way voice channels and provision for the standard rescue relay.

Successive generations began with the launch of four Inmarsat-2 satellites built by Matra Marconi in a consortium led by British Aerospace and launched between October 1990 and April 1992, two by Delta rocket and two by Ariane 4. As demand increased and the capabilities grew the satellites got bigger in response to the requirement and to the evolution of satellite technology. The most recent addition, Inmarsat-6 satellites weigh 5,470kg (12,060lb) and have a capacity for handling 8,000 channels and 200 spot beams, the first being launched in December 2021, the second in February 2023 and both using electric propulsion to reach geostationary orbit.

Inmarsat has evolved over the decades and now carries emergency communications for maritime rescue services free and available throughout the world as part of the Global Maritime Distress and Safety System. The organisation outgrew the Intelsat model and in April 1999 it was split into the International Mobile Satellite Organization (IMSO), which relates to global safety and security, and into Inmarsat Ltd, the operating company for the satellite constellations. At the end of May 2023, the US mega-corporation Viasat bought Inmarsat Ltd for a cash settlement of $850million.

The universality of satellite communication across land and sea served as a platform for rescuing endangered people anywhere on Earth. The system works by

BELOW • *Built by SSL, this Echostar represents the evolving generation of telecommunications satellites carrying data, voice, digital relay signals and TV. (SSL)*

ABOVE • *Launched on December 7, 1966, NASA's experimental communications satellite ATS-1 was one of several satellites used to conduct the first worldwide TV broadcast on June 25, 1967, in which the fate and future of the world was themed. (HAC)*

piggy-backing small relay devices on satellites launched by a range of operators and for different purposes which is responsive to emergency beacons carried by aircraft, ships and individuals sending an emergency signal at 406MHz. The entire operation is managed by the International Cospas-Sarsat Programme, subscribed to by more than 45 countries with distress alerts directed to more than 200 states on a no-cost basis so that national emergency services can swing into action.

After the initiative had been launched by the US, Canada, France, and Russia in 1979, the first emergency occurred on September 10, 1982. By the end of 2021 Cospas-Sarsat had responded to 17,663 events, rescuing 57,413 people, of which 69 per cent were maritime emergencies, 23 per cent were on land and 11 per cent involved aircraft. The remainder were from a variety of miscellaneous platforms. The shift toward low-Earth orbit broadband and satellite communications systems secures an advantage, providing a stronger connection between rescue services and the relayed beacon signal.

Culture wars

Throughout the history of communication satellites, the single most outstanding broadcasting event of historic significance was the Our World programme, the first live multinational, multi-satellite TV broadcast. Organised in the UK by the BBC and with 14 countries as a two-hour event screened on June 25, 1967, it was a landmark event impossible to achieve without TV relay satellites and gathered audiences around the globe. Linked by a network including Early Bird, two Intelsat II and NASA's ATS-1, it involved nine ground stations and 43 control rooms connecting North America, Europe, Tunisia, Japan, and Australia.

Each country had its own anchor, Cliff Michelmore at the BBC in England and James Dibble hosting the US segment. It took 10 months and 10,000 technicians to put it all together with politicians and national leaders barred from the screen. It was an outpouring of deep connection to the planet with separate sections presenting dire warnings about over-population, an appeal for physical fitness and healthy living for the individual, an appeal for global artistic expression among all cultures and a projection to the future showing Cape Canaveral, radio telescopes and pictures of the then most distant galaxy before returning to London.

As the Vietnam War was escalating, the Beatles were asked to perform a finale, presenting the debut apolitical performance of 'All You Need is Love', specially written for the programme to evoke a message of unity conveyed to an audience of 700 million people in 25 countries, more than 15 per cent of the world's population. It was released later as a single. Periodically, Our World is quietly remembered by those who worked the programme, those on the receiving end including many who were deeply moved by its message and some who left the country of their birth to work for what they perceived to be a better future.

The US government used for propaganda the Beatles song which became an anthem for young protesters in communist countries evoking an expression of freedom away from autocratic oppression. Paradoxically, it was also used by campaigners in America protesting about the war in Vietnam and to this day is the poster-song for pacifists as well as some less well-defined organisations on the fringes of society. Few perhaps know where it all began – with satellite television and a politics-free expression of global concern for a rapidly changing planet.

It had knock-on effects when the Beatles tried to set up a commune on the shores of the Aegean Sea based on the principles of the British writer Aldous Huxley. It changed the cultural identity of many people and is credited with having accelerated the social transformation of the young by creating a sense that they were disenfranchised by a world totally out of synch with growing feelings of concern for the future.

Our World was arguably the most influential event in the history of satellite broadcasting. Many believe that, connected to developments in the early days of environmental awareness, concerns about climate change and the growth of international communications it had a

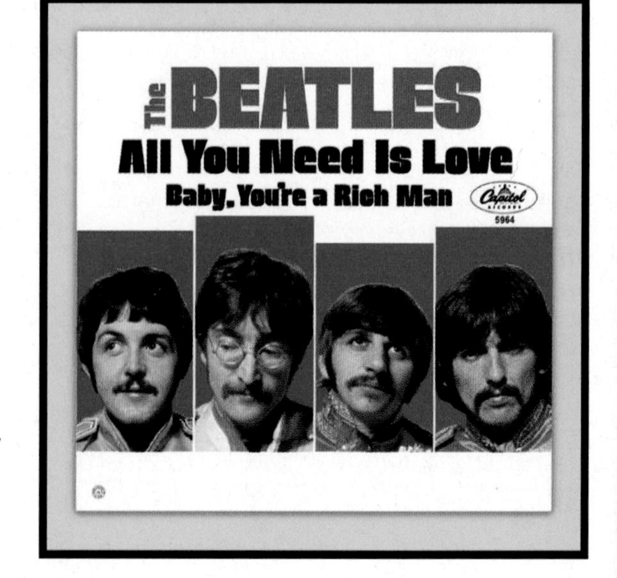

RIGHT • *Seeking an apolitical theme in which all politicians were banned from appearing, the 'Our World' TV broadcast commissioned The Beatles to write a song for the finale which became one of their most famous lyrics of all time. (Capitol Records)*

seminal influence on the way public opinion could change the priorities of politicians. All the more so since it came two years before the first Moon landing, by which time less than half the population supported the Apollo programme.

It also fuelled opposition to the techno-future proclaimed by so many as a panacea for expanding population growth, environmental crises over which concern grew large in the 1970s and which resulted in the redirection of effort within the space programme, discussed in previous chapters. Thus seeded, many of these issues would rise to attract political support for Earth science programmes such as Landsat and the gradual shift by government agencies to set up policies through organs such as ESSA and NOAA and legislation to ban toxins destroying the ozone layer.

In contrast, the constellation of global telecommunications satellites was ready in July 1969 for 650 million worldwide viewers to watch the first Moon landing and then for the first TV images from the surface as Neil Armstrong made his first steps on the surface. There is a causal link between the idealism of the Our World programme and the futurist message of Apollo 11 just two years later, where both served a common global audience with different stimuli, creating a deep concern for environment as related in an earlier chapter and a disenchantment with super-technology.

The unification and convergence of the two messages came with the realisation that to fully define the changing environment, satellites were an essential part of creating a data base for action while leaving it to the global population and its representatives to act upon that. In both respects, satellites were key to both a technological and a cultural transformation in both human action and deliberations about the future.

BELOW • The Beatles' All You Need is Love *became the theme for peace protests across a generation in the Civil Rights movement and at anti-Vietnam War marches due directly to the 'Our World' broadcast. (Author's collection)*

SATELLITES FOR ALL

Liberation, but satellites eliminate the diversity of ways societies traditionally used to survive.

RIGHT • The need for a dedicated US military communications system began with the IDCSP series, small satellites launched in a cluster and dispersed in orbit. (USAF)

BELOW • Satellite applications are wide and diverse and extend to archaeological surveys such as the extensive mapping of the El Mirador site in Guatemala, now recognised as the largest pyramid complex on Earth, with numerous great cities more than 2,500 years old. (Dennis Jarvis)

While civilian and commercial communication satellite systems have dominated the global networks, military command and control has bought heavily into the Space Age. While security organisations and government agencies use spy satellites to know what's going on over the next hill, the everyday management and operation of national defence forces relies on military communication satellites.

Between the launch of Sputnik 1 in October 1957 and the establishment of NASA a year later, the Advanced Research Projects Agency took control of the US Army's Advent programme, highly ambitious and involving an anticipated 10 satellites, each weighing 544kg (1,200lb), with the last six placed in geostationary orbit. A newly formed Advent Management Agency was set up at Fort Monmouth, New Jersey. Advent would have been launched by Atlas-Agena and eventually by Atlas-Centaur.

Work on Advent began in late1958 and was approved in April 1960, but the technical challenges were too great, and it was cancelled two years later, replaced by the Initial Defense Communications Satellite Program (IDCSP) by defence secretary Robert McNamara. It was to be the first in a maturing sequence of military communications satellites under the auspices of the Defense Communications Agency (DCA), which had been set up in 1960.

IDCSP envisaged 24-30 satellites placed at 33,915km (21,075 miles), just below geostationary orbit so that they would slowly drift 27.8° per day. Over a 4.5-day period, they would be visible to a fixed ground station before disappearing over the horizon, by which time the next would be above the opposite horizon and itself visible for four and a half days. These were very small satellites, each a 26-sided polygon with a diameter of 86cm

(33.8in), weighing 45kg (99lb) and supporting a single 3.5W transponder receiving at 8.025GHz and transmitting at 7.25-7.3GHz.

IDCSP satellites were launched in clusters by the powerful Titan IIIC, the first flight of seven on June 16, 1966, with communications linking the US, England, and Germany. With

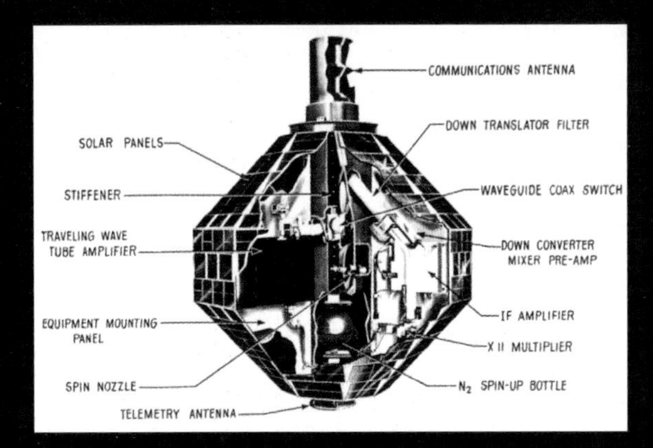

FAR LEFT • The tiny 45kg (99lb) IDCSP satellites were launched between June 1966 and June 1968 and provided an experimental constellation for planning a more permanent and capable system. (USAF)

LEFT • The Advent military satellite programme was initially undertaken by ARPA but was cancelled and replaced with a programme of technology development leading to an operational system. (ARPA)

seven satellites, the second group lifted off on August 26, but the launcher suffered a failure, and they were lost. The third constellation, launched on January 18, 1967, successfully placed eight satellites in orbit, with a third set of four on July 1 before the last eight went up on June 13, 1968. These last two groups were renamed DSCS-I after launch.

New capabilities

Although limited to ground stations equipped with very large antennas, the system demonstrated new capabilities, including the relay of aerial reconnaissance photographs from distant locations back to the Pentagon. This had particular value during the Vietnam War and pushed demand for a more capable system, which appeared in spectacular fashion in a declaration of intent from Robert McNamara, unexpectedly announcing the DSCS-II programme.

Built by TRW, each weighed around 562kg (1,238lb) and consisted of a cylindrical tub-shaped structure with solar cells around the circumference producing 535W of electrical power, two large narrow-beam steerable dish antennas on top and two 20W travelling wave tube amplifiers (TWTAs). A TWTA is a vacuum tube that amplifies radio-frequency signals in the microwave range. Each satellite had 500MHz bandwidth and could handle 1,300 voice channels or data at 100mbps.

This was a significant advance on the small IDSCS/DSCS-I satellites and allowed the use of ground stations with antennas only 2.4m (8ft) in diameter. The last four satellites had 40W amplifiers and with a conflict half a world away driving a need for advanced communications, the DSCS-II satellites were enablers that filtered new and innovative capabilities across all land, sea, and air services. The first was launched on November 3, 1971, to a fixed geostationary (non-drifting) orbit and the last of 16 went up on October 30, 1982, by which date the third generation was being launched.

The Army also wanted a different type of communication satellite, one designed to support tests with ships, aircraft, tanks and jeeps and any static object. Produced by Hughes Aircraft Company, TACSAT was a spin-stabilised satellite with a de-spun antenna platform which, by rotating in the opposite direction would enable the antennas to point in a fixed direction. It had a height of 3.4m (11.15ft) and a diameter of 3m (10ft) with a weight of 725kg (1,598lb).

Launched by a Titan IIIC on February 9, 1969, TACSAT could provide a usable communication system to 0.3m (1ft) diameter antennas and establish links with small vehicles, submarines at sea and aircraft on operations. This was the largest and most powerful satellite launched thus far and it demonstrated outstanding technical capabilities and operating performance, finally being

BELOW • The DSCS series of military communication satellites proved beneficial to global military command and control infrastructure and were extensively trialled during the Vietnam War. (USAF)

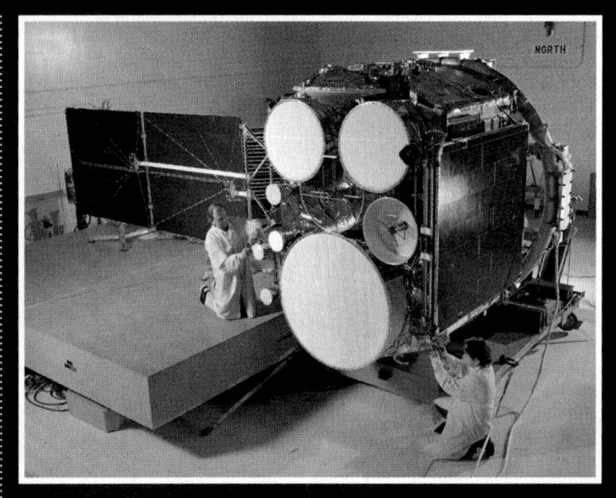

RIGHT • Integration and test of DSCS satellites in a series that were launched from 1971 to 2003. (USAF)

retired in December 1972 and its mission integrated with future generations of DSCS satellites.

The DSCS-III series were hardened against radiation to give some measure of protection in the event the Soviet Union detonated a thermonuclear bomb in space. The electromagnetic pulse (EMP) was a formidable threat to all satellites and to evaluate that, the telecommunications equipment was exposed to nuclear weapon tests on the ground. The Pentagon was preparing for a very different type of threat and the DSCS-III satellites carried a much greater technical performance. They were designed for a hardened, anti-jam environment supporting defence communications at the highest levels around the world and with the White House.

Each satellite had six channels compared with two for the DSCS-II, two 40W and four 10W. The programme spurred development of solid-state power amplifiers which benefitted the entire civil and military telecommunications industry and improvements were fed into successive satellites. By the end of the programme, each satellite was equipped with four 16W amplifiers and two 40W TWTAs. As built by General Electric and Hughes, each satellite weighed 1,235kg (2,723lb) and 14 were launched between October 1982 and August 2003.

To serve both tactical and strategic requirements, Milstar satellites are an advanced, secure, anti-jam and radiation-resistant constellation, of which six were launched between February 1994 and April 2003. With a weight of 4,500kg (9,900lb) and built by Lockheed Martin, each satellite involved elements provided by TRW, now part of Northrop Grumman, and Hughes, now owned by Boeing. A powerful demonstration of new and emerging technologies, in 2010 the programme was handed over to the Pentagon's Advanced Extremely High Frequency (AEHF) office.

Six launches

AEHF satellites are built by Lockheed Martin and Northrop Grumman, each weighs 9,000kg (20,000lb) at launch and 6,168kg (13,598lb) after propellant is consumed for establishing its geostationary orbit. Six AEHF satellites have been launched to date, the first in August 2010 but that failed when the on-board propulsion system malfunctioned. The other five were successfully launched between May 2012 and March 2020, a constellation of global geosynchronous satellites which evolved and integrated several separate functions previously served by individual programmes.

Numerous countries have developed their own military communications satellite programmes but the first outside the US to exploit the geostationary slot for defence use was Britain with its Skynet programme. To date, five generations of Skynet have been launched, the first in November 1969. The two Skynet 1 satellites were built by Philco Ford in the US, but Marconi Space Systems produced Skynet 2, followed by British Aerospace for Skynet 4.

The initial requirement for the Skynet series had been influenced by the British presence east of Suez but as that changed, the requirement for Skynet 3 became redundant and the more pressing requirement to support NATO and European defence interests defined successive generations of Skynet satellite. The most recent Skynet 5, the last of four in the series, was launched in December 2012, with Skynet 6 scheduled for launch in 2025 on a SpaceX Falcon 9. Previously, beginning with the second Skynet 4 launched in December 1988, the UK had used the European Ariane launch vehicles.

Getting there

Communication has been a vital ingredient for going places – be it across town, to the next country, to another continent or across to other worlds. It connects people to information, about their destination, their travel conditions, how to go and what to use to get there. Telephones, mobile phones, and satellite communications have been enabling technologies for most people on the planet and they continue to grow in capacity and affordability. But they carry information from one place to another and that itself enables connectivity linking data to the end user – you and me. It allows travellers to know where they are and where they are going, redefining maps, and adding additional information to explore the journey.

Maps having been fundamental to getting around anywhere on Earth for more than 2,500 years. The drive to improve the way coastlines, hills, mountain ranges and pathways are laid out for reference has been a strong force in creating a record of where we are and where we are going. But that is difficult when there are no maps, or

BELOW • The Milstar military communications programme produced six launches between 1994 and 2003 for worldwide, jam-resistant connections with high levels of encrypting for exchanging classified materials. (USAF)

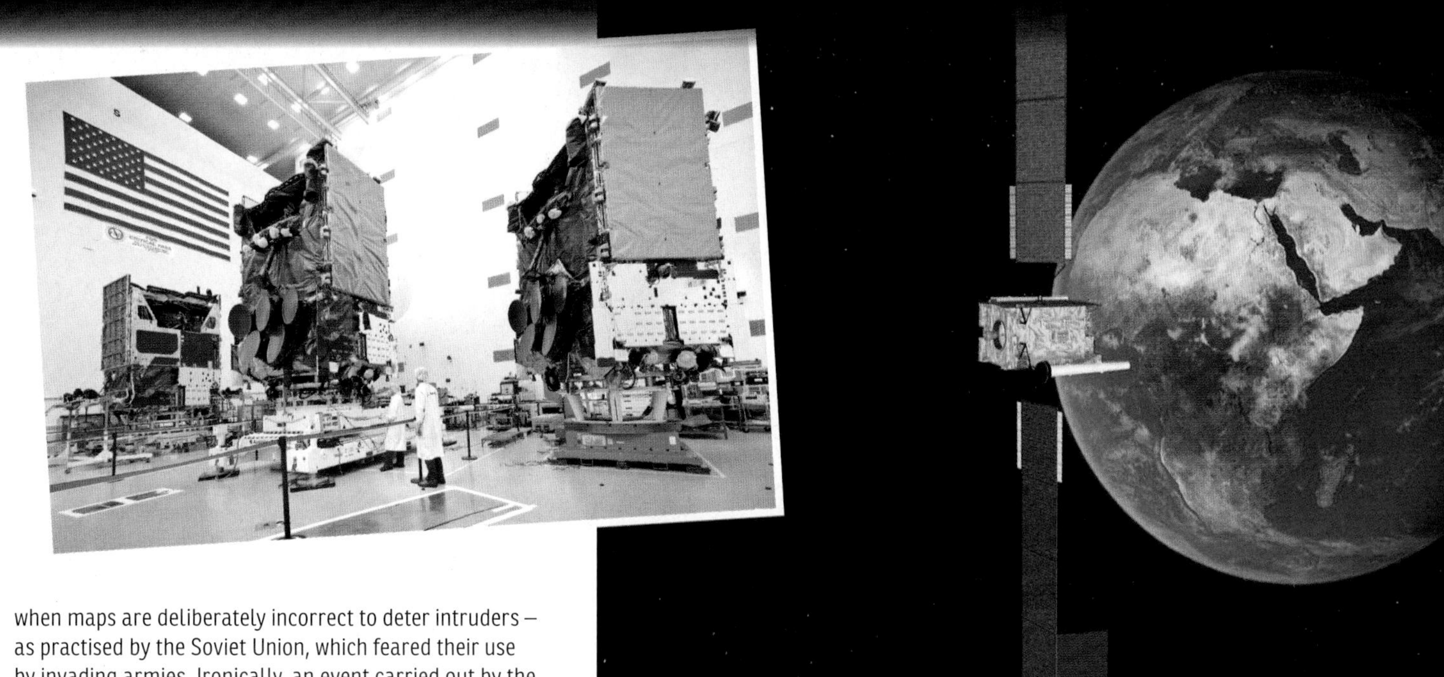

when maps are deliberately incorrect to deter intruders – as practised by the Soviet Union, which feared their use by invading armies. Ironically, an event carried out by the USSR broke through the veil of obfuscation.

It was Sputnik 1, launched in October 1957, that inspired William Guier and George Weiffenback from the Johns Hopkins University Applied Physics Laboratory to begin monitoring the signal and to use the Doppler effect to calculate successive orbits with pencils and slide rules. From that, it was a mere step to conversely use the signal from a satellite to locate the position of an observer. Which was highly relevant for the US Navy as they sought ways to determine the geodetic position of their Polaris missile-launching submarines.

Dr F T McClure persuaded the Johns Hopkins University to develop a satellite navigation system, from which came the Transit programme. Using the accurate position in the sky of three operational and three back-up satellites, the precise location of a ground receiving station such as a submarine could be obtained by using the Doppler principle. This resulted in a military research programme and NASA's Invention Award being given to Dr McClure on January 17, 1961.

Named Transit 1A, the first satellite was launched on September 17, 1959, a 122kg (270lb) sphere with a diameter of 91.4cm (36in) and a band of solar cells around its circumference. Due to a failure in the upper stage of the Thor-Able launch vehicle, it never reached orbit but operated for 24min before plunging back through the atmosphere, sufficient time however for it to prove the concept would work. The system aimed for a position accuracy of 926m (3,040ft), defined as 0.5 nautical miles, as that was the Navy requirement with an ultimate goal of 185m (608ft), or 0.1nm.

Launched on April 13, 1960, Transit 1B was placed in an orbit of 760km (472 miles) by 382km (237 miles) at an inclination of 51° demonstrating the system as highly effective in the 89 days it operated before a technical failure ended transmissions. A succession of different generations of Transit satellites followed, embraced within the Navy Navigation Satellite System (NNSS), which formed the basis for spin-off programmes such as Oscar, Triad and Timation.

Over time, a worldwide network of tracking stations was set up to operate through a series of geodesic satellites to build a global database for accurate positions around the globe. Thus established, the accuracy of any location would grow as the reference points using the highly accurate position of the navigation satellites were refined with greater precision.

The first of these, ANNA (Army, Navy, and NASA) was launched on May 10, 1962, and marked numerous positions on the Earth's surface and to measure the strength and location of its gravitational field. NASA also contributed through its GEOS series, exclusively for geodesic mapping.

Throughout the 1960s and 1970s, a wide range of different satellite programmes refined and perfected satellite navigation, many flying as piggy-back on other flights going up into similar orbits. A vast amount of data was acquired on the effects of solar and cosmic radiation on the satellites, their precise location in space, the effects of the outer atmosphere on signals and the requirements for achieving highly accurate timing devices which were vital to the concept of satellite navigation. Throughout the early 1970s, the ability to build a fully operational system took root in Project 621B, with outstanding results.

Project 621B became the Global Positioning System (GPS) which, together with the telephone and the worldwide web is perhaps the most significant technology touching the lives of more people around the globe than any other invention. The GPS system was designed to provide information on longitude, latitude, altitude, speed, and time through a constellation of 24 satellites orbiting the Earth in 12-hour orbits. Each satellite had a highly stable atomic clock, which were synchronised together and corrected by ground stations for any drift.

ABOVE • *The Skynet programme gave the UK the first military communications net outside the US and Russia, with launches from 1971, the first two satellites being procured from the US with all since 1974 being manufactured in the UK. (UKMOD)*

ABOVE LEFT • *Satellites of the Wideband Global Satcom series were launched from 2007 through to the present day. (USAF)*

LEFT • *Here being displayed by Maj Alex Pestrichella (left) in 1968, the Vela satellite which was launched to detect nuclear tests and employed to provide early warning of new and clandestine entrants to the nuclear club. (USAF)*

RIGHT • At Schriever Air Force Base, California, control of the Navstar GPS satellites is carried out by US Air and Space Force personnel. (USSAF)

Distance measuring

The system works by measuring the distance between the point of transmission in space and its reception on the ground, using the speed of light as a constant value. The time on both the satellite and the ground station is synchronised. A wide range of stations collect data so that the satellites' locations in space are accurately determined. Correlation between the exact position of the satellites and their synchronised clocks and with the ground is vital.

Four satellites are optimum, by which a GPS receiver can decode the signal and make a determination using triangulation. Receivers are passive in that they collect rather than transmit and that makes the system very useful to the military user, who does not need to emit a traceable signal that could give away the location. It also makes the system unique to the operator because it can be encoded so that no unauthorised user is able to receive signals and determine their precise location.

Going under the name Navstar and built by Rockwell International, the first of 11 Block I GPS satellites was launched on February 22, 1978, the last on October 9, 1985. The system proved to have teething problems, but all were solvable and Navstar was declared operational in 1989. Manufactured by Lockheed Martin and later Boeing, the Block II series followed with the first launch on February 14, 1989 and the last of 59 on February 5, 2016, although there were several derivative evolutions in the same series. The first of the Lockheed Martin Block III series went into orbit on December 23, 2018 and the latest of six of that type went up on January 18, 2023.

A significant breakthrough which enabled greater progress with Navstar, Code Division Multiple Access (CDMA) enabled four-dimensional positioning available to the user without a necessity for the receiver, hand-held or otherwise, to have an atomic clock. Precision applications are only possible for any user anywhere around the globe because of CDMA and that has been adopted for telecommunications satellites, adding greater potential for stacking up more users on the same frequencies.

BELOW • Beginning in the 1970s, development of the Navstar GPS navigation system has expanded beyond all expectation with both civil and military applications throughout the world. (USAF)

It follows the earlier use of Time Division Multiple Access (TDMA) and Frequency Division Multiple Access (FDMA), which attaches several users to the same time slot or frequency and switches in microseconds the specific part of the band for separate users. For instance, by splitting one second of transmission into, say 10 users, each one gets 0.1sec of time but the human brain interprets the conversation as an uninterrupted flow and ignores the gaps. In precisely the same way, a film strip presents a run of separate static frames which the eye interprets as a flow of uninterrupted movie action.

Over the last 45 years in which the GPS satellites have been up and running, the technology has changed markedly. The satellites have become bigger and heavier, and their reliability has improved. The Block I series had a weight of 759kg (1,674lb), increasing to 1,660kg (3,660lb) for the initial Block II, with some sub-variants weighing up to 2,032kg (4,480lb). Some Block III variants weigh 4,400kg (9,700lb) and have universal connectivity so that an update command to one is shared by all. Navstar satellites orbit the planet in a 12-hour orbit at a distance of 20,200km (12,550 miles) and typically at an inclination of 55°.

In total, 78 Navstar satellites have been launched, 31 are presently operational and 42 have been retired. Two were launch failures and three were either unhealthy or operated as spares. The advancing stages of GPS technology opened a wide range of commercial applications. Although developed as a system for the military, in 1973 President Ronald Reagan opened a civilian use of GPS for improved safety with commercial aviation, but with limited accuracy which was integrated with the Block II satellites as a separate channel.

In 1999, Benefon released a GPS telephone, and the technology began to show up in cars and trucks. The following year, three additional channels were added for civilian use and that enhanced accuracy ten-fold. The price of processing chips for GPS receivers dropped from around $3,000 to £1.50 and that led to an exponential growth for a very wide range of services including in-car navigation, location-based services, personal use in watches and mobile phones and for ships, railways, aircraft, and a wide diversity of industries.

These include farmers using GPS for precision location of field boundaries and a new generation of self-drive agricultural vehicles, rescue boats automatically supplied with the optimum route to a disaster based on sea current, and remote controlled unmanned aerial vehicles (UAVs) for exploring remote areas and uninhabited regions. They have also introduced the remotely controlled Unmanned Combat Air Vehicle (UCAV) operated by personnel half a planet away, spying or dropping GPS-guided munitions. There is even a family of GPS-guided tank shells.

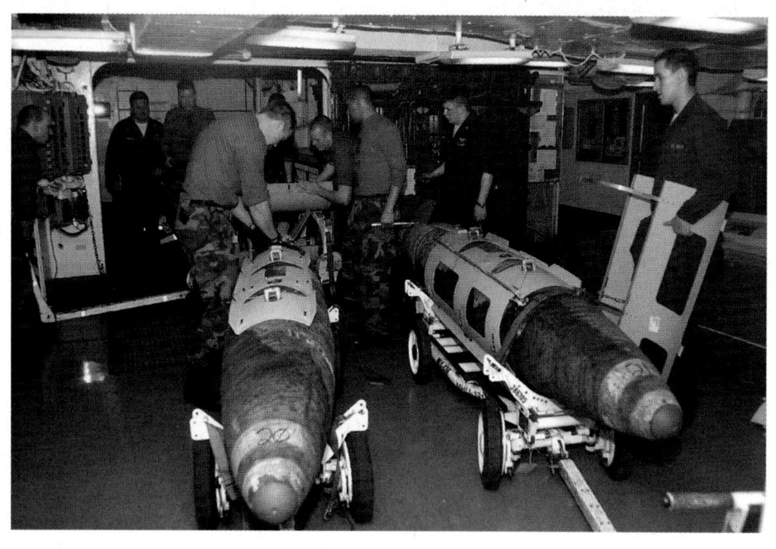

ABOVE • *A GPS receiver package is attached to a Joint Direct Attack Munitions (JDAM) bomb for in-flight targeting and high accuracy. GPS also allows artillery and tank shells to be guided to their targets. (USAF)*

ABOVE LEFT • *The Defense Support Program (DSP) early-warning satellites were launched between 1970 and 2007, replaced now by the SBIRS satellites. (USAF)*

Saturation point?

Since 1980, GPS applications have generated almost $4.14trillion in economic benefits and saved several thousand lives through avoiding natural disasters by navigating around catastrophic events such as volcanic eruptions. In the US, at least 52 billion gallons of fuel have been saved, cutting one trillion vehicle miles through improved route planning and car or truck navigation. The scientific study of the Earth and its environment is enhanced as several thousand unmanned surface boats only a few metres across float freely along ocean currents, their position known accurately by GPS trackers and data sent back by communication satellites.

The Navstar GPS system has competitors with the BeiDou system from China, Galileo from the European Union, GLONASS from Russia, NavIC from India and QZSS from Japan. The expanding range of operators is collectively known as the Global Navigation Satellite System (GNSS) with specific capabilities supporting priority services. These allow politicians to integrate national objectives and new applications are emerging, including road-pricing schemes using in-car trackers to record routes by cars and trucks for billing users according to a charging tariff, loaded for congestion, exhaust emissions or speed limits.

Given suitable equipment installed in road vehicles, it would be possible to control the speed of a car by activating an engine limiter based on a GPS locator. An attempt in this direction was quietly sought by the administration of UK prime minister Tony Blair in the early 2000s when it pushed for the European Galileo system to integrate a locked-in road pricing capability with an option for speed control through the GNSS network. At the time, it was too ambitious, and the European motor industry quietly sloughed it off. That has returned to further consideration supporting optional emissions regulations and the attraction of replacing fuel taxation lost to electric vehicles with a tax on road use and speed controls.

Today, there are more than seven billion GNSS tracking devices in use and that is expected to grow to 10.5 billion by 2031, far outnumbering the human population of eight billion. Of the current total, almost 80 per cent of smartphone owners use satellite navigation applications, with Google Maps accounting for two-thirds of those. These applications embrace many niche areas such as archaeology, marine biology, animal, and bird tracking, as well as walking and running routes integrated with GNSS and smartphones. Both use local cell-phone towers for providing their services, as do in-car navigation systems provided by mapping organisations that overlay GNSS signals on to the road network. If your car directs you up a blind alley it's not the fault of the satellite but rather the map provider.

Communication satellites and the burgeoning use of remote-sensing satellites from a growing host of providers using small satellites launched at low cost is changing the world at a dizzying pace. With satellite navigation, the integration of services into a common platform is already well underway, driven largely by the needs of disaster-relief and the media industry bringing timely and accurate news of conflict.

News and information have never been more instant and complete, no longer biased by political propaganda or subjective reporting. In war zones, remote-controlled unmanned aerial vehicles carry GNSS, digital cameras, satellite communication links and electronic tagging. Every human casualty, armoured vehicle and tank can be located, photographed, and instantly identified. It is the same with natural disasters where information about conditions on land or at sea are known instantly, the position of every ship at sea or aircraft anywhere around the world can be instantly identified and tracked.

Reliance on space

In the 21st century, the vast majority of the Earth's population obtains its weather reports from space, receives news from around the world on TV screens fed from platforms in geostationary orbit and gets directions for

BELOW • *Navstar satellites in production with their service now universally supported by an expanding industry of applications for commercial and personal use. (USAF)*

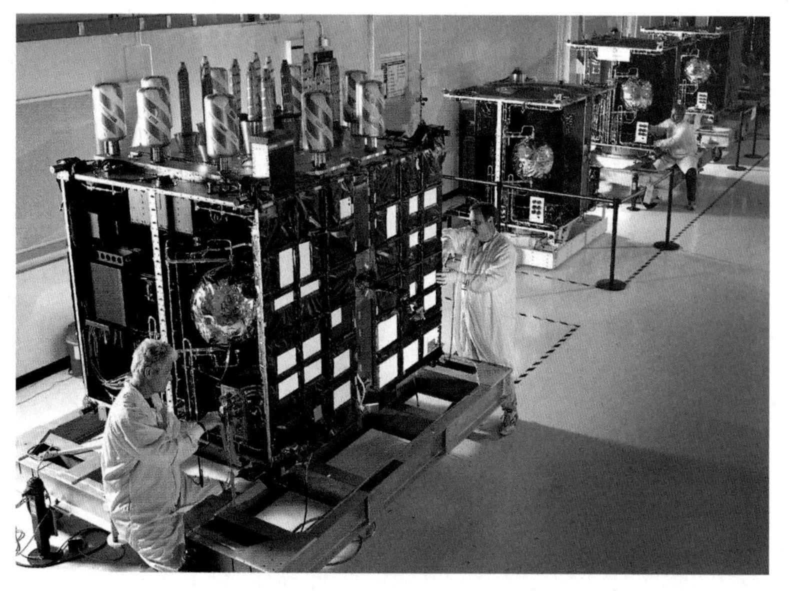

travel via the global navigation systems. Environmentalists, conservationists, town and rural planning officials and countless specialist organisations also get information directly from satellites.

Around the world, banks and financial institutions work at millisecond intervals to manage and control money for vast international conglomerates and individuals alike – by satellite. And national security relies on spies in space, robot eyes that look in visual and non-visual parts of the spectrum to monitor threats. But it comes at a price.

Within a very few years, the number of satellites in orbit will reach an unsustainable level of density due to the commercial clamour for more services to more people. National agencies keep account of all objects in space, down to very small scales. The European Space Agency estimates that in near-Earth space, around the altitude in which the International Space Station resides, there are more than 36,000 objects larger than 10cm (4in) across, with a million at 1-10cm (0.4-4in) and possibly in excess of 130 million particles smaller than 1cm.

Large structures such as the ISS can manoeuvre if there is warning of a potentially harmful object heading for it and now some small satellites are capable of manoeuvring out of the way. But the very act of sending things into space is accompanied by lots of debris from severed stages, ejected fixtures, and mountings, even the odd nut or bolt! But while the very act of launching a satellite has to have a licence to do that, and communications satellites must file an application for a set of frequencies it wants to use at a specified location over the equator, there is little or no control on who does what in space itself.

At the beginning of 2022, there were 8,200 satellites in orbit, of which almost 4,900 were still operating. Around 12 per cent were in geosynchronous or geostationary orbit, three per cent were in medium-Earth orbit and 1.2 per cent were in highly elliptical orbit. But a staggering 83 per cent were in low-Earth orbit, where most of the danger lies from close encounters, near collisions and the not-so-occasional direct impact.

By May 2023, with 4,300 in orbit, SpaceX is the largest

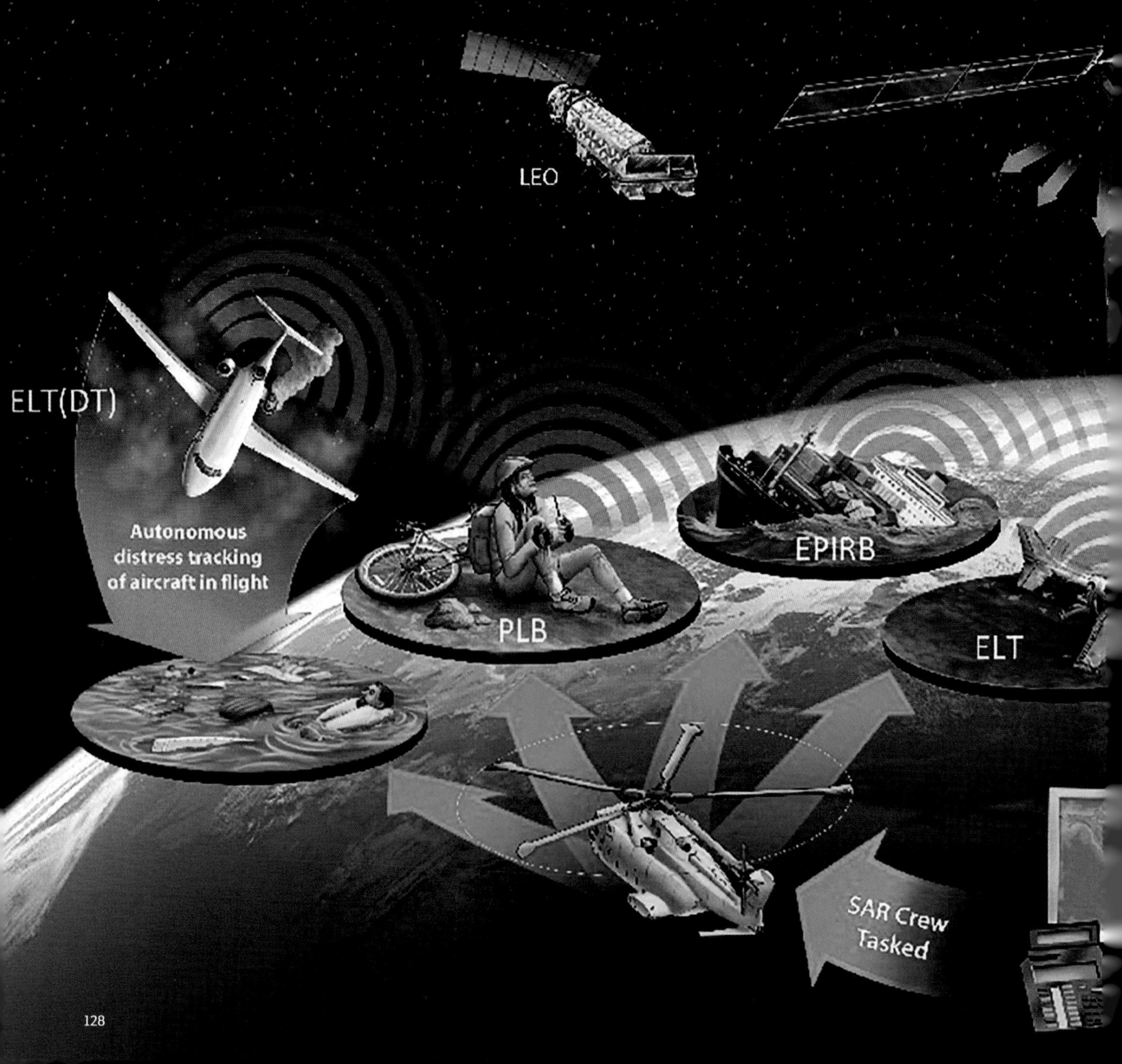

BELOW • An enduring legacy of the prolific array of satellites in space, the Cospas/Sarsat system relays distress calls from people in peril to local and regional rescue organisations. Many lives have been saved by this passive system. (Cospas/Sarsat)

satellite operator on the planet, followed by China with more than 470 and OneWeb with almost 300. The US military has about 70, NASA has just over 60 and the European Space Agency has more than 50. A large number of private operators have the rest. SpaceX has a declared intention of getting a total of 42,000 of its Starlink communication satellites in orbit by the early 2030s. Added to which are around 4,000 from broadband connector OneWeb and 3,200 from Amazon for a similar service. China has already declared plans for 13,000 of its Guowang system.

The threat from collisions and impacting debris is a major concern for space operators around the world, as it always has been. Added to which is the vulnerability of a new industrial age based not on diversity of supply but on the single supply of electrical power, the only means by which our new capabilities can be provided.

Perhaps the greatest challenge for all is not the proliferation of satellites and the vital services they perform but rather the infrastructure of a space-based world we are not equipped to survive in the event of its collapse. Satellites have liberated humans, but they have also eliminated the diversity of ways that societies have traditionally used to survive, through diversity and optional ways to find solutions. We are in a new world order managed solely through systems we cannot access on a regular, day-to-day basis and only control with signals we send into space.

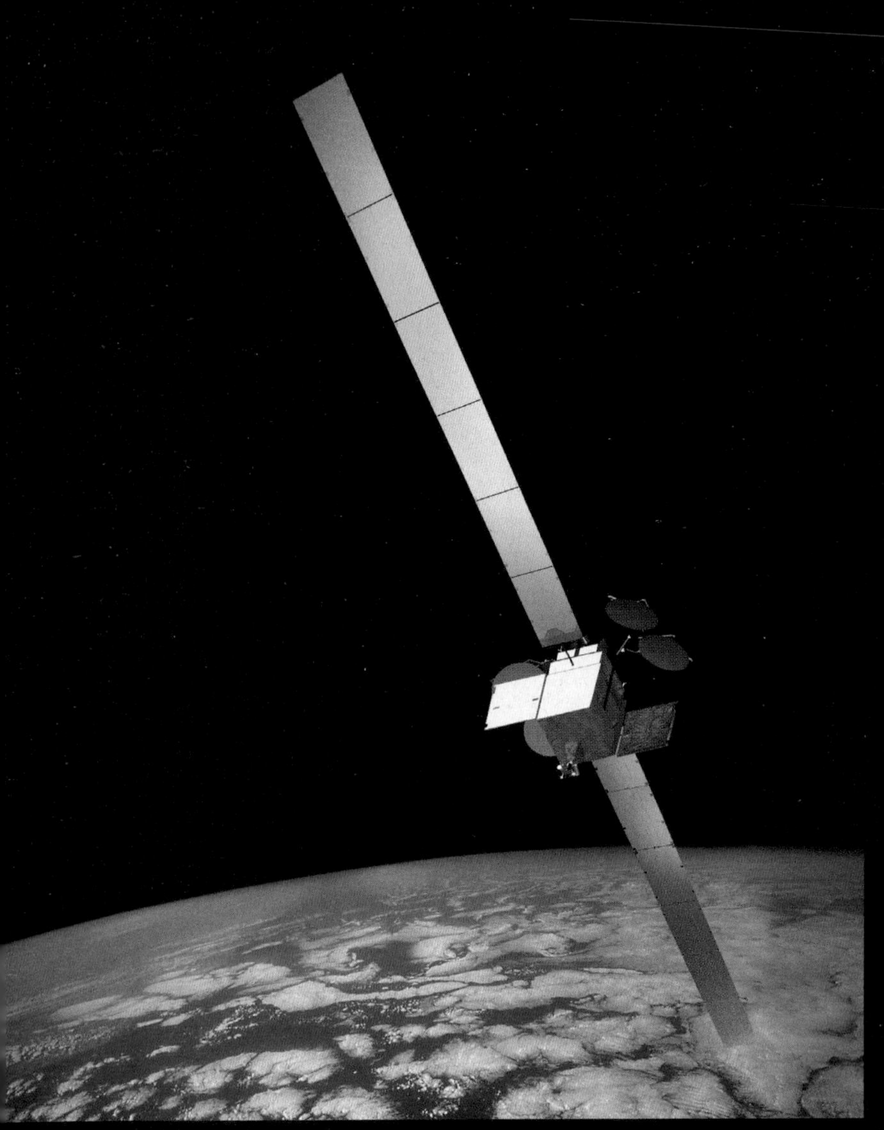

SPACECRAFT AND SATELLITES GLOSSARY

ABOVE • ***(US Air and Space Force)***

A-4	German Army designation for the V2 rocket
APT	Automatic Picture Transmission
ARDC	Air Research and Development Command
ARPA	Advanced Research Projects Agency
ASLV	Augmented Satellite Launch Vehicle
ATS	Applications Technology Satellite
AT&T	American Telephone and Telegraph
CIA	Central Intelligence Agency of the United States
CNES	Centre National d'Études Spatiales, France's national space agency
CNSA	China National Space Agency
CST	Commercial Space Technologies
DMSP	Defense Meteorology Satellite Program
DOI	Department of the Interior of the United States
ELDO	European Launcher Development Organisation
EOSAT	Earth Observation Satellite Company
ERTS	Earth Resource Technology Satellite
ESRO	European Space Research Organisation
ESSA	Environmental Science Services Administration
GE	General Electric
Geostationary orbit	One in which the satellite is located in the plane of the equator and takes 24 hours to go once around the Earth

Geosynchronous orbit. One in which the satellite takes 24 hours to complete one orbit of the Earth

GOES	Geostationary Operational Environmental Satellite
GPS	Global Positioning System
GSLV	Geosynchronous Satellite Launch Vehicle
Hz	Hertz or cycles per second of a radio frequency
ICBM	Intercontinental Ballistic Missile
IGY	International Geophysical Year (1957-58)
ISAS	Institute of Space and Astronautical Science of Japan
JAXA	Japan Aerospace Exploration Agency
Kármán Line	Height at which international agreement defines the start of space – 100km (62 miles)
KH	Key Hole classification of spy satellite optical system
LACIE	Large Area Crop Inventory Experiment
LC	Launch Complex
LIDAR	Light Detection and Ranging
Mach	The speed of sound named after Ernst Mach
MMS	Multi-Mission Spacecraft
MOL	Manned Orbiting Laboratory
MSS	Multispectral Scanner
MT	Megaton equivalent explosive yield of a nuclear weapon
NACA	National Advisory Committee for Aeronautics
NASA	National Aeronautics and Space Administration
NASDA	National Space Development Agency of Japan
NOAA	National Oceanic and Atmospheric Administration
NRO	National Reconnaissance Office
NSC	National Security Council
OMB	Office of Management and Budget

Polar orbit. A path around the Earth at 90° to the equator which passes over both North and South Poles

PSLV	Polar Satellite Launch Vehicle
R-1	Russia's first ballistic missile
R-2A	Russian sounding rocket
R-7	Russian first-generation ICBM
RBV	Return Beam Vidicon
RCA	Radio Corporation of America
SCI	Space Consultants International

Sounding rocket. Rocket used for ballistic flights with instruments for scientific measurements

SLBM	Submarine Launched Ballistic Missile
SLV	Satellite Launch Vehicle
SPOT	Systéme Probatoire d'Observation de la Terre
SRBM	Short Range Ballistic Missile
SRV	Satellite Recovery Vehicle
SSO	Sun-synchronous orbit in which the satellite is slightly offset from a true polar orbit, generally at 98° to the equator so that it tracks the same sun angle as the Earth moves in its orbit

TDRSS. Tracking and Data Relay Satellite System

U-2	Lockheed spy plane
USDA	United States Department of Agriculture
USGS	United States Geological Survey
V1	German pulse-jet powered glide bomb of World War Two
V2	German ballistic missile of World War Two
WAC	Upper stage for V2 rocket